高等职业教育系列教材

COMPUTER TECHNOLOGY

HarmonyOS
应用开发与实践

主编｜季云峰　李涛　高云
参编｜刘　丽　徐威　匡亮　平震宇

本书以物联网智慧农业移动端应用系统的开发为例，讲解了在 HarmonyOS 中进行移动开发的方法，将开发过程拆解成 14 个任务，模拟类似敏捷开发的流程，任务的设计既遵循 HarmonyOS 知识点的学习路径，又尽量符合移动应用开发的教学过程。通过增量迭代来开发各个功能模块，读者完成每个任务后都可以编译、运行，并且每个任务可以在前一任务的基础上进行拓展，最终完成整个应用的开发。读者可通过重构、复原该系统来掌握物联网应用开发的常用技术。同时，本书还引入了 Git 版本管理系统，可帮助读者熟悉企业开发的过程。

本书适合作为高职院校电子信息类、计算机类相关专业的教材，也适合对物联网移动应用开发感兴趣的读者阅读学习。

本书配有授课电子课件，需要的教师可以登录www.cmpedu.com免费注册，审核通过后下载，或联系编辑索取（微信：13261377872，电话：010-88379739）。

图书在版编目（CIP）数据

HarmonyOS 应用开发与实践 / 季云峰，李涛，高云主编. —北京：机械工业出版社，2022.12（2024.8 重印）
高等职业教育系列教材
ISBN 978-7-111-71881-9

Ⅰ. ①H… Ⅱ. ①季… ②李… ③高… Ⅲ. ①移动终端-应用程序-程序设计-高等职业教育-教材 Ⅳ. ①TN929.53

中国版本图书馆 CIP 数据核字（2022）第 198153 号

机械工业出版社（北京市百万庄大街 22 号　邮政编码 100037）
策划编辑：王海霞　　责任编辑：王海霞
责任校对：张艳霞　　责任印制：李　昂

北京中科印刷有限公司印刷

2024 年 8 月第 1 版·第 2 次印刷
184mm×260mm·14.25 印张·371 千字
标准书号：ISBN 978-7-111-71881-9
定价：59.00 元

电话服务　　　　　　　　　网络服务
客服电话：010-88361066　　机 工 官 网：www.cmpbook.com
　　　　　010-88379833　　机 工 官 博：weibo.com/cmp1952
　　　　　010-68326294　　金 书 网：www.golden-book.com
封底无防伪标均为盗版　　　机工教育服务网：www.cmpedu.com

二维码清单

名称	图形	名称	图形
0-1　任务探究		3-1　任务探究	
1-1　任务探究		3-2　创建 Splash 界面	
1-2　安装 DevEco Studio		3-3　实现 Splash 倒计功能	
1-3　创建 HelloWorld 工程		3-4　优化 Splash 界面功能	
1-4　修改 App 图标和名称		4-1　任务探究	
2-1　任务探究		4-2　创建新大陆云平台项目	
2-2　安装 Git		4-3　创建物联网行业实训仿真项目	
2-3　实践 Git 基本功能		5-1　任务探究	
2-4　实践 Git 标签管理		5-2　安装 Postman，并测试百度地址	
2-5　将 SmartAgriculture 工程加入 Git 版本管理		5-3　调试新大陆云平台——用户登录 API	

名称	图形	名称	图形
6-1　任务探究 1		8-3　绑定 TabList 与 PageSlider	
6-2　任务探究 2		9-1　任务探究	
6-3　设计登录界面		9-2　创建土壤、控制界面，并实现我的界面	
6-4　实现登录界面登录按钮逻辑		9-3　创建云平台参数设置功能	
6-5　使用兼葭库实现登录云平台功能		10-1　任务探究	
7-1　任务探究 1		10-2　持久化账号、云平台等相关参数	
7-2　任务探究 2		11-1　任务探究 1	
7-3　LoginAbilitySlice（登录界面）切换到 MainAbilitySlice（首页）		11-2　任务探究 2	
7-4　实现底部导航栏功能		11-3　实现兼葭拦截器打印日志	
8-1　任务探究		11-4　全局使用 HiLog 日志	
8-2　创建大气环境监控界面		11-5　从云平台获取传感器数据	

二维码清单

名称	图形	名称	图形
12-1 任务探究		14-2 任务探究 2	
12-2 创建土壤监控		14-3 全屏显示	
13-1 任务探究		14-4 退出当前账户	
13-2 创建执行器控制		图 0-2 智慧农业监控系统的硬件拓扑结构	
14-1 任务探究 1 多语言设计			

Preface 前 言

2019 年 8 月 9 日，在华为开发者大会上，HarmonyOS（鸿蒙操作系统）正式发布，并实行开源，自此，面向未来的国产操作系统正式诞生。为适应技术的未来发展趋势和国产化浪潮，本书尝试基于 HarmonyOS 开发一个物联网智慧农业应用系统，来覆盖整个移动应用开发课程的教学过程。

本书依据移动应用开发岗位能力需求，基于职业工作过程、模块化课程设置和项目化教学实施的需要，融入移动应用开发技能大赛和职业技能等级证书要求，通过将该系统分解成 14 个任务来支持项目化、模块化教学的需求。每个任务都设定了知识目标和技能目标，实现了对 HarmonyOS 应用开发知识点和技能点的覆盖，并将"构建国产自主可控的信息系统"这一科技强国理念融入教学全过程。

系统使用华为的 DevEco Studio 工具来开发，模拟类似敏捷开发的流程，通过增量迭代来开发各个功能模块，读者完成每个任务后都可以编译、部署、运行应用，可以直观地看到学习效果，每个任务都在前一任务的基础上进行拓展，最终完成整个系统的开发。

本书同时引入了 Git 版本管理系统，帮助读者提前培养正规的开发流程和习惯，熟悉企业开发的过程。本书的代码同时托管于 Gitee 平台上，每个任务都打上了标签，读者可以根据具体章节下载对应的代码，每个标签的代码都是完整、可以运行的，且与任务化、项目化一一对应。

本书是一本介绍物联网方向 HarmonyOS 移动应用开发的教材，因此在教材中对部分常用的 HarmonyOS 开发知识点没有介绍，如分布式、服务卡片、流转、多编程语言开发等，读者可以根据自己的需求进行拓展学习。

本书由江苏信息职业技术学院季云峰、李涛、高云、刘丽、徐威、匡亮、平震宇编写，季云峰、李涛、高云担任主编，刘丽、徐威、匡亮、平震宇参与编写。

由于编者水平有限，编写时间仓促，尽管我们尽了最大的努力，但书中仍难免有不妥和错误之处，恳请读者批评指正。

编　者

目 录 Contents

二维码清单

前言

绪论 智慧农业项目概述及设计 ································ 1

0.1 项目背景 ································ 1
0.2 项目方案 ································ 1
0.3 系统部署 ································ 2
0.4 系统功能 ································ 3
0.4.1 Splash 欢迎界面 ················ 4
0.4.2 系统登录界面 ···················· 4
0.4.3 大气环境监控界面 ·············· 4
0.4.4 土壤环境监控界面 ·············· 4
0.4.5 水阀控制界面 ···················· 5
0.4.6 参数设置界面 ···················· 6

任务 1 开发环境搭建和创建工程 ································ 7

1.1 初识 HarmonyOS ························ 7
1.2 HarmonyOS 平台架构 ·················· 7
   1.2.1 HarmonyOS 的三大特征 ········ 8
   1.2.2 HarmonyOS 系统架构 ··········· 8
1.3 创建开发环境和 HelloWorld 工程 ································ 10
   1.3.1 开发准备 ·························· 10
1.3.2 开发应用 ·························· 10
1.3.3 运行应用 ·························· 12
1.3.4 发布应用 ·························· 13
1.4 更改应用的启动图标和应用名称 ·· 13
   1.4.1 复制 logo.png 图片 ············· 13
   1.4.2 配置图标 ·························· 13
   1.4.3 配置应用名称 ···················· 14

任务 2 认识 Git 版本管理 ································ 16

2.1 版本管理 ································ 16
   2.1.1 本地版本管理系统 ·············· 16
   2.1.2 集中化版本管理系统 ·········· 16
   2.1.3 分布式版本管理系统 ·········· 17
2.2 Git 的初识与安装 ······················ 17
2.2.1 Git 的安装 ························ 18
2.2.2 Git 快照流 ························ 18
2.2.3 Git 的三个区 ···················· 19
2.2.4 Git 基本配置 ···················· 19
2.3 实践 Git 基本功能 ···················· 20

VII

2.3.1	初始化 Git 仓库 20	2.6.1	分支创建 29
2.3.2	查看当前状态 21	2.6.2	分支切换 31
2.3.3	增加文件 21	2.6.3	分支合并 32
2.3.4	查看 log 24	2.7	Git 标签管理 32

2.4　Git 版本回退 24

- 2.4.1　commit id 24
- 2.4.2　reset 命令 25

2.5　Git 内容对比 26

- 2.5.1　对比工作目录与暂存区 26
- 2.5.2　对比仓库不同版本 27
- 2.5.3　对比工作目录与仓库 28
- 2.5.4　对比暂存区与仓库 28

2.6　Git 分支与查看 29

- 2.7.1　查看标签 32
- 2.7.2　创建标签 32
- 2.7.3　指定版本打标签 33
- 2.7.4　检出标签 34
- 2.7.5　删除标签 35

2.8　将工程加入 Git 版本控制 35

- 2.8.1　初始化工作目录 35
- 2.8.2　提交代码到本地仓库 38
- 2.8.3　将该版本代码打上标签 40

任务 3　创建 Splash 界面 41

3.1　HarmonyOS 应用的基础知识 41

- 3.1.1　用户应用程序 42
- 3.1.2　用户应用程序包结构 42
- 3.1.3　关键术语 43

3.2　HarmonyOS 应用的配置文件 43

- 3.2.1　配置文件的组成 44
- 3.2.2　配置文件的元素 44
- 3.2.3　配置文件内部结构 45
- 3.2.4　app 对象的内部结构 45
- 3.2.5　deviceConfig 对象的内部结构 46
- 3.2.6　module 对象的内部结构 48

3.3　HarmonyOS 应用的资源文件 54

- 3.3.1　resources 目录 54
- 3.3.2　限定词目录 55
- 3.3.3　资源组目录 56

3.4　创建 Splash 界面 56

- 3.4.1　了解项目工程 56
- 3.4.2　了解 Ability 基础 57
- 3.4.3　创建 Splash Ability 和布局 58
- 3.4.4　编辑配置文件 59
- 3.4.5　编辑 Splash 布局 61
- 3.4.6　编辑 Splash Ability 66

3.5　提交代码到仓库 73

任务 4　创建新大陆云平台"智慧农业"项目 76

4.1　创建云平台项目 76

- 4.1.1　了解新大陆物联网云平台 77
- 4.1.2　创建新大陆物联网云平台"智慧草坪"项目 77
- 4.1.3　创建传感器 79
- 4.1.4　创建执行器 79

4.2　创建物联网行业实训仿真项目 80

4.2.1	创建仿真项目	80	4.4	使用新大陆 1+X 传感网设备设计原型 … 82
4.2.2	调试智慧农业数据采集和控制	80	4.4.1	认识新大陆 1+X 传感网设备体系 … 82
4.3	使用新大陆物联网云平台数据模拟器 … 81		4.4.2	搭建感知层原型 … 82

任务 5　RESTful API 调试 … 84

5.1	HTTP 基础 … 84		5.3	使用 Postman 调试 API 接口 … 89
5.1.1	HTTP 消息结构	84	5.3.1	Postman 安装 … 89
5.1.2	HTTP 方法	86	5.3.2	Postman 基本使用 … 90
5.1.3	HTTP 常用方法 GET 和 POST	86	5.4	调试新大陆物联网云平台 API 接口 … 91
5.1.4	HTTP 常见请求头部	87	5.4.1	归纳新大陆物联网云平台 RESTful API … 91
5.1.5	HTTP 常见响应报头	87	5.4.2	调试用户登录 API … 92
5.1.6	HTTP 状态码	88	5.4.3	查询设备最新数据 … 94
5.2	RESTful 架构 … 88		5.4.4	模糊查询传感器 … 101
5.2.1	REST 概述	88	5.4.5	发送命令控制设备 … 103
5.2.2	资源与 URI	88		
5.2.3	统一资源接口	89		

任务 6　创建登录功能 … 106

6.1	编辑登录界面 ability_login.xml … 106		6.3.3	增加网络权限和 HTTP 访问 … 115
6.2	编辑登录逻辑 LoginAbilitySlice.java … 111		6.4	登录云平台 … 115
			6.4.1	创建 Wan 接口 … 116
6.3	引入网络库蒹葭（JianJia）… 114		6.4.2	创建 Account Bean … 116
6.3.1	添加 mavenCentral()仓库	114	6.4.3	登录逻辑 … 117
6.3.2	添加依赖	114	6.4.4	编译运行 … 120
			6.5	提交代码到仓库 … 120

任务 7　创建底部标签导航栏 … 121

7.1	不同 Page Ability 的切换 … 121		7.1.2	根据 Ability 的全称启动应用 … 122
7.1.1	掌握 Intent 意图	122	7.1.3	根据 Operation 的其他属性启动应用 … 123

IX

7.1.4	LoginAbilitySlice 切换到		7.2.2	StackLayout	128
MainAbilitySlice	124	7.2.3	ScrollView	128
7.1.5	编译运行	125	7.2.4	TabList	128
7.1.6	提交代码到仓库	126	7.2.5	实现 TabList 功能	130
7.2	使用 TabList 设置多标签	126	7.2.6	编译运行	133
7.2.1	Component	126	7.3	提交代码到仓库	134

任务 8 创建大气环境监控界面 …………… 135

- 8.1 使用 PageSlider 组件切换页面 …… 135
  - 8.1.1 增加 PageSlider …… 136
  - 8.1.2 创建 PageSliderProvider 子类 …… 136
- 8.2 大气监控界面设计 …… 137
- 8.3 更新 MainAbilitySlice.java …… 138
  - 8.3.1 PageSlider 常用方法 …… 138
  - 8.3.2 更新 MainAbilitySlice.java 代码 …… 139
  - 8.3.3 编译运行 …… 142
- 8.4 提交代码到仓库 …… 142

任务 9 创建参数设置界面 …………… 143

- 9.1 个人设置界面 …… 143
  - 9.1.1 创建土壤界面 …… 144
  - 9.1.2 创建控制界面 …… 144
  - 9.1.3 创建我的界面 …… 144
  - 9.1.4 编辑 MainAbilitySlice.java …… 148
  - 9.1.5 编译运行 …… 149
  - 9.1.6 提交代码到仓库 …… 149
- 9.2 云平台参数设置界面 …… 149
  - 9.2.1 创建云平台参数设置界面 …… 149
  - 9.2.2 创建 AbilitySlice 的 Java 文件 …… 152
  - 9.2.3 更新 MainAbilitySlice.java …… 153
  - 9.2.4 编译运行 …… 154
- 9.3 提交代码到仓库 …… 154

任务 10 参数持久化 …………… 155

- 10.1 AbilityPackage 类 …… 155
- 10.2 轻量级数据存储 …… 155
  - 10.2.1 轻量级数据存储概述 …… 155
  - 10.2.2 轻量级数据存储开发 …… 156
- 10.3 更新 Java 代码 …… 160
  - 10.3.1 更新 MyApplication.java 文件 …… 160
  - 10.3.2 更新 CloudParameterSettingAbilitySlice.java …… 167
  - 10.3.3 更新 SplashAbilitySlice.java …… 171

10.3.4	更新 LoginAbilitySlice.java	171
10.4	编译运行	173
10.5	提交代码到仓库	173

任务 11 从云平台获取传感器数据 174

11.1	设置兼葭（JianJia）拦截器	174
11.1.1	兼葭（JianJia）拦截器	175
11.1.2	更新 MyApplication.java 文件	175
11.1.3	编译运行	176
11.1.4	提交代码到仓库	176
11.2	使用 HiLog 日志	176
11.2.1	HiLog 日志基础	177
11.2.2	更新项目代码	178
11.2.3	编译运行	180
11.2.4	提交代码到仓库	180
11.3	从云平台获取传感器数据	180
11.3.1	更新 MyApplication.java	180
11.3.2	创建 SensorData.java	181
11.3.3	更新 Wan.java	182
11.3.4	更新 MainAbilitySlice.java	183
11.3.5	编译运行	188
11.3.6	调试解决 Bug	188
11.4	提交代码到仓库	189

任务 12 创建土壤监控界面 190

12.1	更新 pageslider_soil.xml 界面	190
12.2	更新 MainAbilitySlice.java 代码	190
12.3	更新 string.json	193
12.4	编译运行	193
12.5	提交代码到仓库	194

任务 13 创建执行器控制 195

13.1	使用 Switch 组件	195
13.2	更新 pageslider_control.xml 文件	196
13.3	更新 java 文件	200
13.3.1	创建 CmdRsp.java bean 文件	200
13.3.2	更新 Wan.java	200
13.3.3	更新 MainAbilitySlice.Java	201
13.4	更新 string.json 文件	205
13.5	编译运行	205
13.5.1	打开水阀	205
13.5.2	关闭水阀	206
13.5.3	设备未上线	206
13.6	提交代码到仓库	206

任务 14 创建多语言环境 207

14.1 多语言设计 207
14.2 全屏显示 209
14.3 退出当前账号 210
  14.3.1 更新 MyApplication.java 文件 210
  14.3.2 更新 SplashAbilitySlice.java 文件 211
  14.3.3 更新 LoginAbilitySlice.java 文件 211
  14.3.4 更新 SplashAbility.java 文件 212
  14.3.5 更新 MainAbilitySlice.java 文件 212
  14.3.6 了解 Page Ability 生命周期 213
  14.3.7 编译运行 215
14.4 提交代码到仓库 215

参考文献 216

绪论　智慧农业项目概述及设计

任务概述

本任务主要介绍 HarmonyOS（鸿蒙系统）物联网移动应用系统的设计，了解智慧农业的项目背景、项目方案、系统部署，以及系统的主要功能。

知识目标

- 了解物联网系统设计。
- 了解物联网云平台。
- 了解敏捷开发。
- 了解物联网系统部署。
- 了解智慧农业任务线。

技能目标

- 能绘制智慧农业物联网系统。
- 能绘制智慧农业物联网系统数据流。

0.1　项目背景

物联网智慧农业监控系统移动端应用是用于监控管理农业的 HarmonyOS 应用项目。通过该系统可以利用智能移动终端实时感知农业领域的大气、土壤等环境信息，同时可以在智能终端实时控制农业领域的水阀系统。

0.2　项目方案

智慧农业物联网系统如图 0-1 所示。感知层的数据传递到云平台，手机端的 App 可以和云平台进行交互，获取感知层传感器的数据，以及控制感知层的设备。感知层可以通过新大陆公司的物联网行业实训仿真软件进行虚拟设计，也可以通过实物搭建真实的环境，或者直接在云平台端使用数据模拟器生成测试数据。手机端可以采用 Android、HarmonyOS 和 iOS 平台。本项目采用 HarmonyOS 平台进行 App 设计。

图 0-1　智慧农业物联网系统

如果感知层采用虚拟仿真技术来搭建，则智慧农业监控系统的硬件拓扑结构如图 0-2 所示。该系统的主要硬件如下。

1）光照传感器、温湿度传感器、PM2.5 传感器、大气压力传感器、土壤水分传感器、风速传感器等接入 ZigBee（一种短距离、低速率的无线通信技术）模块，实现光照度、温湿度、PM2.5 值、大气压力、土壤水分、风速等模拟数据的采集。

2）风向传感器、CO_2 传感器接入 ADAM-4017 采集器，实现对风向、CO_2 浓度等信息的采集，继电器接入 ADAM-4150，实现对风扇开关的控制。

3）ADAM-4150 通过 485 总线接入物联网数据采集网关，各 ZigBee 节点通过 ZigBee 网络接入到物联网数据采集网关。

4）物联网数据采集网关通过 Wi-Fi 接入互联网，连接到部署在公网的物联网云平台。

5）移动端设备通过物联网云平台实现远程控制和管理。

6）项目中以四个风扇来代替水阀。

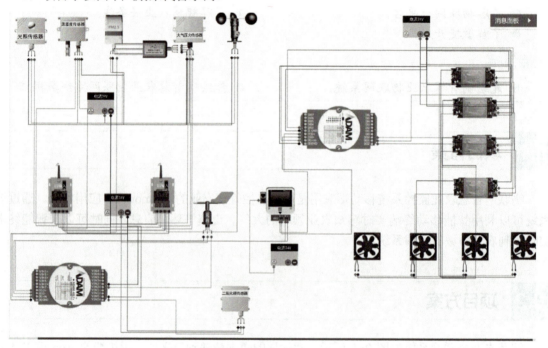

图 0-2　智慧农业监控系统的硬件拓扑结构

0.3　系统部署

读者可以基于新大陆实训设备（NLE-JS2000）和私有云进行软硬件系统的安装和部署，也可以使用新大陆虚拟仿真软件，利用新大陆免费开放的公网物联网云平台（http://www.nlecloud.com）完成系统的搭建。系统安装部署步骤如下。

1）烧写和组网 ZigBee 节点（或 LoRa 节点、NB-IoT 节点），连接各硬件设备。

2）部署和配置物联网云平台（私有云）。

3）配置物联网数据采集网关。

4）部署智能终端 App。

0.4 系统功能

系统功能包含欢迎界面、登录界面、大气环境监控界面、土壤环境监控界面、水阀控制界面和参数设置界面。通过 14 个任务的学习，完成整个 App 的开发，如图 0-3 所示。

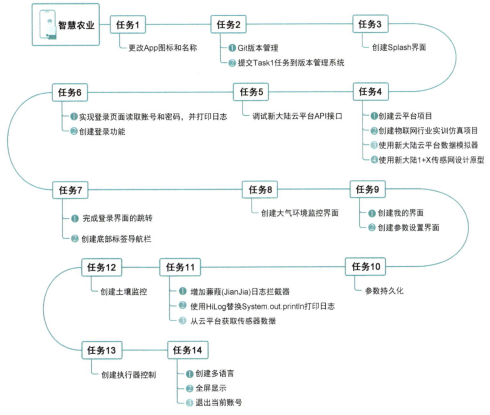

图 0-3 智慧农业任务线

后面的 14 个任务采用类似敏捷开发的方式，每个版本完成一个任务，即完成一个功能，通过增量迭代来开发各个模块，逐步增加功能，最终完成整个 App 的功能，智慧农业敏捷开发如图 0-4 所示，引入 Git 版本管理，与图 0-3 相对应，每完成一个完整的任务，就给 App 打上一个标记（Tag），代表当前标记的版本完成了既定功能，后期想要提取之前标记的版本，可以通过输入代码"git checkout tag_name"来提取目标版本进行测试。

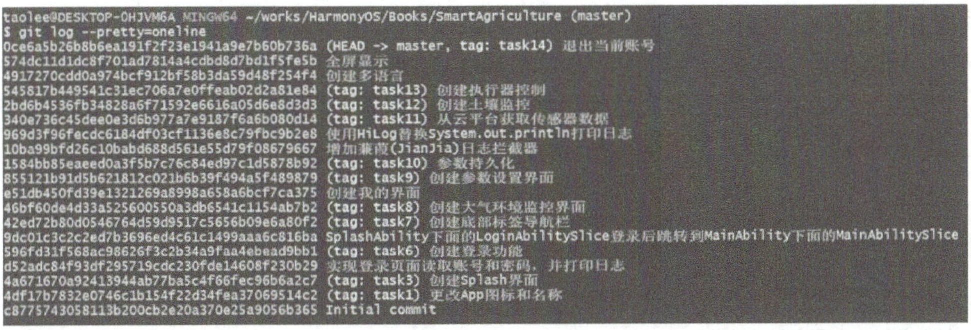

图 0-4 智慧农业敏捷开发

下面分别介绍各个界面的功能。

0.4.1　Splash 欢迎界面

系统的 Splash 欢迎界面展示了 App 的 Logo 与 App 的名称"智慧农业",如图 0-5 所示。欢迎界面展现时自动倒计时 6s,当倒计时到达 6s 或者提前单击"取消",则进入下一个页面。

0.4.2　系统登录界面

系统提供了用户登录功能,登录界面如图 0-6 所示,用户通过该页面输入手机号和密码,单击登录,则 App 自动访问新大陆云平台,在新大陆云平台端进行账号验证,验证通过后,返回加密字符串(Token)给 App 端,而后 App 进入下一个界面。

图 0-5　欢迎界面

图 0-6　登录界面

0.4.3　大气环境监控界面

系统主界面实时展现大气环境监控数据,如图 0-7 所示,该界面提供了大气环境中,温度、湿度、风速、风向、光照、气压、PM2.5 和 CO_2 监控数据,其中数据部分是模拟数据。在页面的下端显示了本次数据同步的时间。在主界面下部提供底部导航栏,可以由此导航到"大气""土壤""控制""我的"界面。

0.4.4　土壤环境监控界面

土壤环境监控界面,显示了土壤环境中的 pH 值、雨量、温度和湿度信息,如图 0-8 所示,其中数据部分是模拟数据。

图 0-7　大气环境监控界面

图 0-8　土壤环境监控界面

0.4.5　水阀控制界面

水阀控制执行器界面显示了 4 个控制器，它们用来控制 4 个水阀，如图 0-9 所示。本系统中，感知层采用风扇代替水阀，打开水阀，即打开感知层的风扇，关闭水阀，即关闭感知层的风扇。如果设备未上线，则会弹窗提示相关设备还未上线，如图 0-10 所示。

图 0-9　水阀控制界面

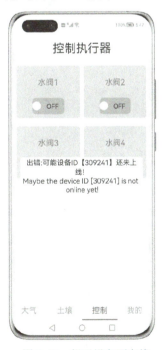

图 0-10　提示设备不在线

0.4.6 参数设置界面

个人设置界面提供账号信息显示,如图 0-11 所示,包含当前登录账号的手机号,AccessToken 信息。同时提供设置功能,包含云平台参数设置,退出当前账号两个功能。单击云平台参数设置,则切换到云平台参数设置界面,如图 0-12 所示,该界面提供设备 ID 展示、大气环境传感器参数设置、土壤环境传感器参数设置、执行器传感器参数设置。

图 0-11 个人设置界面

图 0-12 参数设置界面

任务 1　开发环境搭建和创建工程

任务概述

本任务要完成 HarmonyOS 应用程序开发环境 DevEco Studio 的创建，在 DevEco Studio 中完成 SmartAgriculture 工程项目的创建、配置和运行，并实现应用启动图标和应用名称的修改。

知识目标

- 了解 HarmonyOS API 版本。
- 了解 Gradle 系统构建工具。
- 了解 HarmonyOS 远程虚拟设备。
- 了解 HarmonyOS 本地真机设备。
- 掌握 DevEco Studio 中的工程目录。
- 了解 HarmonyOS 本地虚拟设备。
- 了解 HarmonyOS 远程真机设备。
- 了解 HarmonyOS 应用运行过程。

技能目标

- 能搭建 DevEco Studio 开发环境。
- 能完成 DevEco Studio 工程项目的创建、配置和运行。
- 能完成应用启动图标和名称的修改。

1.1 初识 HarmonyOS

　　HarmonyOS 是华为公司于 2019 年 8 月 9 日至 8 月 11 日在东莞举行的华为开发者大会上正式发布的操作系统。HarmonyOS 是一款面向万物互联时代的、全新的分布式操作系统。HarmonyOS 创造了一个超级虚拟终端互联的世界，将人、设备、场景有机地联系在一起，将消费者在全场景生活中接触的多种智能终端上的应用，实现极速发现、极速连接、硬件互助、资源共享，用合适的设备提供场景体验。

　　HarmonyOS 是华为公司基于开源项目 OpenHarmony 开发的面向多种全场景智能设备的商用版操作系统。

1.2 HarmonyOS 平台架构

　　HarmonyOS 在传统的单设备系统能力基础上，提出了基于同一套系统能力、适配多种终端形态的分布式理念，能够支持手机、平板计算机、智能穿戴、智慧屏、车载计算机等多种终端设备，提供全场景（移动办公、运动健康、社交通信、媒体娱乐等）业务能力。

1.2.1 HarmonyOS 的三大特征

HarmonyOS 提供了支持多种开发语言的 API，供开发者进行应用开发。支持的开发语言包括 Java、XML（Extensible Markup Language）、C/C++、JS（JavaScript）、CSS（Cascading Style Sheets）和 HML（HarmonyOS Markup Language）。该系统具有以下三大特征。

1）搭载该操作系统的设备在系统层面融为一体，形成超级终端，让设备的硬件能力可以弹性扩展，实现设备之间硬件互助，资源共享。对用户而言，HarmonyOS 能够将生活场景中的各类终端进行能力整合，实现不同终端设备之间的快速连接、能力互助、资源共享，匹配合适的设备、提供流畅的全场景体验。

2）面向开发者，实现一次开发，多端部署。对应用开发者而言，HarmonyOS 采用了多种分布式技术，使应用开发与不同终端设备的形态差异无关，从而让开发者能够聚焦上层业务逻辑，更加便捷、高效地开发应用。

3）一套操作系统可以满足不同能力的设备需求，实现统一操作系统，弹性部署。对设备开发者而言，HarmonyOS 采用了组件化的设计方案，可根据设备的资源能力和业务特征灵活裁剪，满足不同形态终端设备对操作系统的要求。

1.2.2 HarmonyOS 系统架构

HarmonyOS 整体遵循分层设计，从下向上依次为：内核层、系统服务层、框架层和应用层。系统功能按照"系统→子系统→功能/模块"逐级展开，在多设备部署场景下，支持根据实际需求裁剪某些非必要的子系统或功能/模块。HarmonyOS 技术架构如图 1-1 所示。

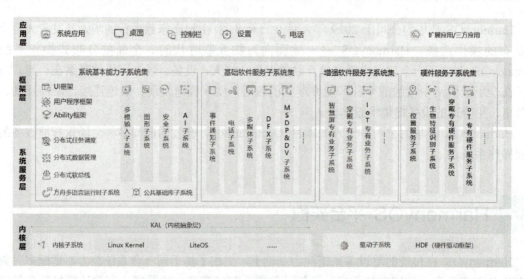

图 1-1 HarmonyOS 技术架构图

1. 内核层

（1）内核子系统

HarmonyOS 采用多内核设计，支持针对不同资源受限设备选用合适的操作系统内核。内核抽象层（KAL，Kernel Abstract Layer）通过屏蔽多内核差异，对上层提供基础的内核能力，包括进程/线程管理、内存管理、文件系统、网络管理和外设管理等。

（2）驱动子系统

硬件驱动框架（HDF）是 HarmonyOS 硬件生态开发的基础，提供统一外设访问能力和驱动开发、管理框架。

2. 系统服务层

系统服务层是 HarmonyOS 的核心能力集合，通过框架层对应用程序提供服务。该层包含以下几个部分。

（1）系统基本能力子系统集

为分布式应用在 HarmonyOS 多设备上的运行、调度、迁移等操作提供了基础能力，由分布式软总线、分布式数据管理、分布式任务调度、方舟多语言运行时、公共基础库、多模输入、图形、安全、AI 等子系统组成。其中，方舟运行时提供了 C 语言、C++、JS 语言的多语言运行时和基础的系统类库，也为使用方舟编译器静态化的 Java 程序（即应用程序或框架层中使用 Java 语言开发的部分）提供运行时。

（2）基础软件服务子系统集

为 HarmonyOS 提供公共的、通用的软件服务，由事件通知、电话、多媒体、DFX（Design For X）、MSDP&DV 等子系统组成。

（3）增强软件服务子系统集

为 HarmonyOS 提供针对不同设备的、差异化的能力增强型软件服务，由智慧屏专有业务、穿戴专有业务、IoT 专有业务等子系统组成。

（4）硬件服务子系统集

为 HarmonyOS 提供硬件服务，由位置服务、生物特征识别、穿戴专有硬件服务、IoT 专有硬件服务等子系统组成。

根据不同设备形态的部署环境，系统服务层中的子系统集内部可以按子系统粒度裁剪，每个子系统内部又可以按功能粒度裁剪。

3. 框架层

框架层为 HarmonyOS 应用开发提供了 Java、C、C++、JS 等多语言的用户程序框架和 Ability 框架，两种 UI 框架（包括适用于 Java 语言的 Java UI 框架、适用于 JS 语言的 JS UI 框架），以及各种软硬件服务对外开放的多语言框架 API（应用程序编程接口）。根据系统的组件化裁剪程度，HarmonyOS 设备支持的 API 也会有所不同。

4. 应用层

应用层包括系统应用和第三方非系统应用。HarmonyOS 的应用由一个或多个 FA（Feature

Ability）或 PA（Particle Ability）组成。其中，FA 有 UI 界面，提供与用户交互的能力；而 PA 无 UI 界面，提供后台运行任务的能力以及统一的数据访问抽象。FA 在进行用户交互时所需的后台数据访问也需要由对应的 PA 提供支撑。基于 FA/PA 开发的应用，能够实现特定的业务功能，支持跨设备调度与分发，为用户提供一致、高效的应用体验。

1.3 创建开发环境和 HelloWorld 工程

开发一个完整的 HarmonyOS 应用需要完成四个流程：开发准备→开发应用→运行应用→发布应用，如图 1-2 所示。

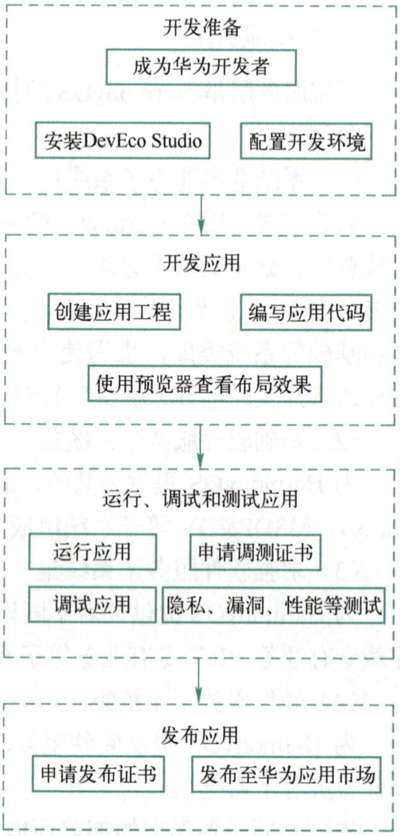

1.3.1 开发准备

在进行 HarmonyOS 应用开发前，开发者需要注册一个华为开发者账号，并完成实名认证。实名认证方式分为"个人实名认证"和"企业实名认证"。

1. 成为华为开发者（个人/企业）

要成为华为开发者要完成以下三步，具体内容请参考代码目录/doc/鸿蒙开发环境搭建.pdf。

（1）注册账号
（2）登录
（3）实名认证

2. 安装 DevEco Studio

具体安装流程参考代码目录/doc/鸿蒙开发环境搭建.pdf。

1.3.2 开发应用

DevEco Studio 集成了 Phone、Tablet、TV、Wearable、LiteWearable 等设备的典型场景模板，可以通过工程向导轻松地创建一个相应模板的新项目。当开发一个 HarmonyOS 应用时，首先需要根据项目创建向导，创建一个新的项目，工具会自动生成对应的代码和资源模板。

双击桌面"DevEco Studio"图标，启动软件，如图 1-3 所示，单击"Create Project"，进入创建新项目界面。

创建项目如图 1-4 所示。①选择模板，此处列了很多模板，默认选中"Empty Ability"模板。②单击"Next"按钮。

图 1-2 HarmonyOS 应用开发流程

图 1-3　DevEco Studio 项目窗口

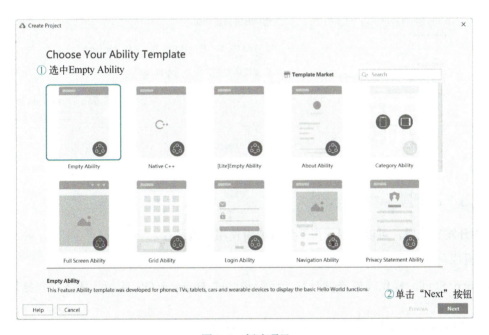

图 1-4　创建项目

进入项目配置阶段，根据向导配置项目的基本信息，如图 1-5 所示。①"Project name"，项目的名称，可以自定义，此处输入 SmartAgriculture。②"Project type"，项目的类型，标识该项目是一个原子化服务（Atom service）或传统方式的、需要安装的应用（Application），此处我们选择"Application"。③"Bundle name"，设置软件包名称，默认情况下，应用 ID 也会使用该名称，应用发布时，应用 ID 需要唯一。此处选择默认设置。④"Save location"，项目文件本地存储路径，路径名不能包含中文字符。⑤"Development mode"，设置开发模式，部分模板支持低代码开发，可选择"Super Visual"，此处我们选择"Traditional coding"。⑥"Language"，选择该项目模板支持的开发语言，可根据模板支持的语言选择 JS 或 eTS 或 Java，其中 eTS 受 DeEco Studio V3.0 Beta2 及以上版本支持，此处选择"Java"。⑦"Compatible API version"，兼容的 SDK 最低版本，此处选择"SDK: API Version 6"。⑧"Device type"，设置该项目模板支持的设备类型，支持多选，默认全部勾选。如果勾选多个设备，表示该原子化服务或传统方式的、需要安装的应用支持部署在多个设备上，此处我们仅勾选"Phone"。⑨以上配置完成后，单击"Finish"按钮，完成工程创建。

图 1-5　配置项目

1.3.3　运行应用

应用开发完成后，可以使用模拟器运行或者使用真机运行。

1. 模拟器

DevEco Studio 提供模拟器供开发者运行和调试 HarmonyOS 应用、服务。包括以下两种方式。

（1）使用本地模拟器（Local Emulator）运行应用

具体内容参考代码目录/doc/鸿蒙开发环境搭建.pdf。

（2）使用远程模拟器（Remote Emulator）运行应用

远程模拟器每次使用时长为 1 小时，到期后会自动释放资源，到期释放后，可以重新申请资源。具体内容参考代码目录/doc/鸿蒙开发环境搭建.pdf。

2. 真机

真机调试包含了远程真机和本地真机，具体内容参考代码目录/doc/鸿蒙开发环境搭建.pdf。

1.3.4 发布应用

HarmonyOS 应用开发一切就绪后，需要将应用发布至华为应用市场，以便分发，普通消费者就可以通过应用市场获取对应的 HarmonyOS 应用。需要注意的是，发布到华为应用市场的 HarmonyOS 应用，必须使用发布证书进行签名。

1.4 更改应用的启动图标和应用名称

创建项目后，可以对应用进行自定义，最简单的就是修改应用的图标和应用的名称。

1.4.1 复制 logo.png 图片

将 logo.png 文件复制到如图 1-6 所示目录，在 DevEco Studio 左侧项目视图中，展开"SmartAgriculture"→"entry"→"src"→"main"→"resources"→"base"→"media"，右击 media 目录并选择"Paste"，即可完成图片的复制。

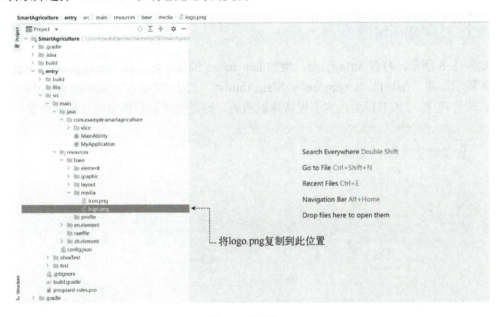

图 1-6　复制 logo

1.4.2 配置图标

双击打开如图 1-7 所示的 config.json 文件，找到""icon": "$media:icon""，将其改为""icon": "$media:logo""，即可完成启动图标的配置。

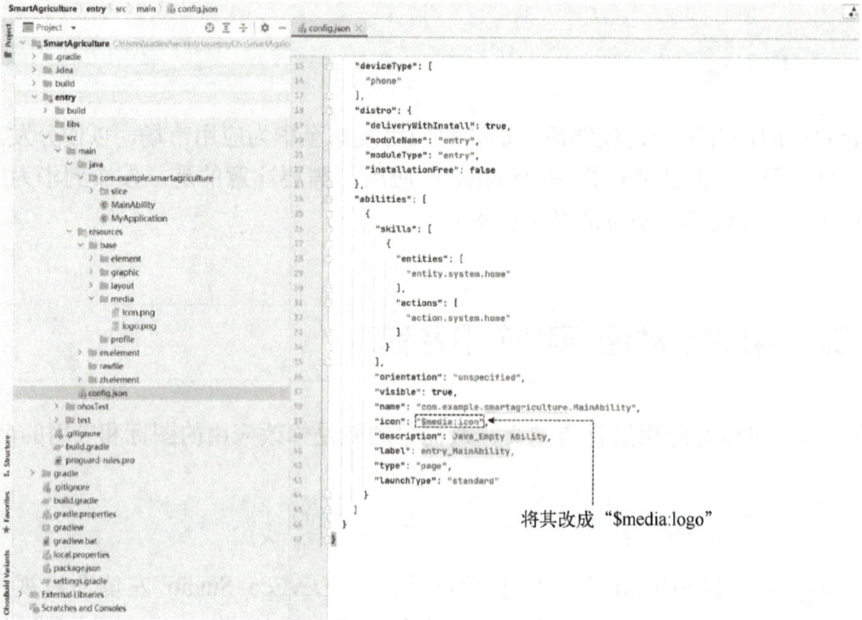

图 1-7 配置 logo

1.4.3 配置应用名称

按如图 1-8 所示，打开 string.json，增加 app_name 的值。然后在 config.json 中修改代码，如图 1-9 所示，将""label": "$string:entry_MainAbility""改为""label": "$string:app_name""。重新编译，运行程序。（本书后面内容不做特殊说明的，则是使用远程模拟器来测试程序）

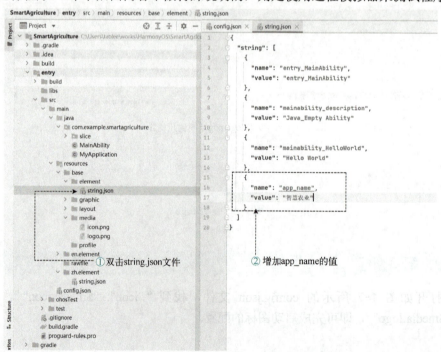

图 1-8 配置 App 名称

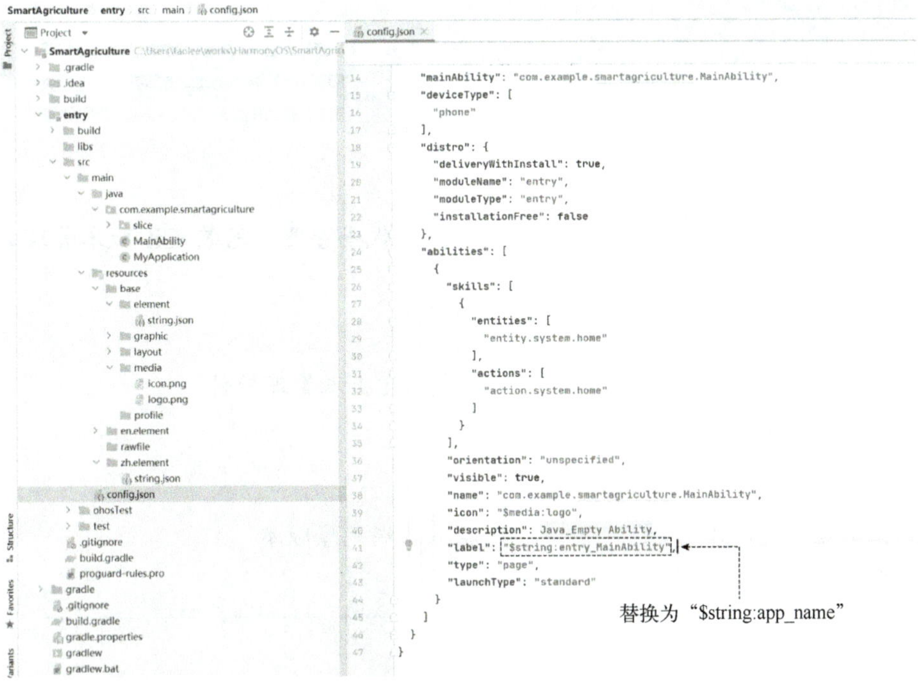

图 1-9 配置 label

如图 1-10 所示，启动图标和应用名称已经完成修改。

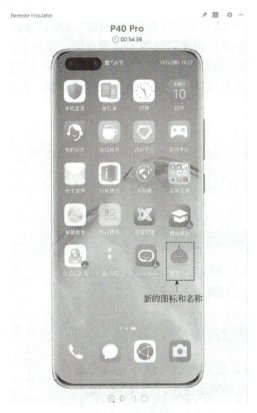

图 1-10 修改后的应用图标和名称

任务 2　认识 Git 版本管理

任务概述

本任务主要学习软件工程领域的重要技能——代码管理。完成 Git 版本管理的学习并将任务 1 的项目代码加入到版本管理系统中。

知识目标

- 了解版本管理的概念。
- 掌握 Git 的基本操作。
- 了解 Git 版本管理软件。

技能目标

- 能在本地构建版本管理。
- 能掌握基本的版本管理命令。
- 能从云端克隆版本。

2.1 版本管理

版本管理是一种记录 1 个或若干个文件内容变化，以便将来查阅特定版本情况的系统。其实，在我们日常工作中已经不经意间使用过版本管理系统，最基本的版本管理如图 2-1 所示，即用不同文件，差异化命名的方式起到管理文件的作用。这种方式缺点很多。如浪费空间，随着版本更新次数越来越多，其占用空间也越来越大。容易出错，而且查看版本不方便。

名称	修改日期	类型	大小
出游计划v1.doc	2021/9/12 21:06	DOC 文档	9 KB
出游计划v2.doc	2021/9/12 21:06	DOC 文档	9 KB
出游计划v3.doc	2021/9/12 21:06	DOC 文档	9 KB
出游计划v4.doc	2021/9/12 21:06	DOC 文档	9 KB

图 2-1　基本版本管理

2.1.1　本地版本管理系统

针对以上问题，有很多本地版本管理系统，采用数据库来记录文件更新的差异，如图 2-2 所示。

2.1.2　集中化版本管理系统

由于本地管理系统的缺点是不善于处理多人协同工作。于是集中化的版本管理系统应运而生，如 CVS 版本管理系统。集中化管理系统有一个单一的集中管理的服务器，它用来保存所有文件的修订版本，而协同工作的人们都通过客户端连到这台服务器，取出

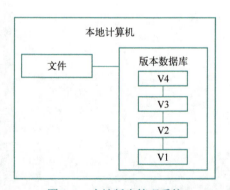

图 2-2　本地版本管理系统

最新的文件或者提交更新，如图 2-3 所示。

2.1.3 分布式版本管理系统

集中化版本管理系统解决了多人协同工作的问题，可以让管理者很方便地查看项目的进度，加快了多人协

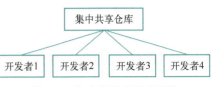

图 2-3　集中化版本管理系统

同工作进展。但是集中化的版本管理系统也存在一些问题，如版本采用集中化管理，一旦服务器出现宕机，则在宕机时间内，任何工程师不能修改提交版本。更严重的是，如果服务器出现磁盘损坏等情况，则会造成整个版本的更新历史丢失。为了解决该问题，分布式版本管理系统出现了，如图 2-4 所示。

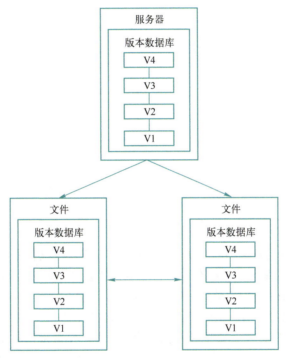

图 2-4　分布式版本管理系统

在分布式版本管理系统中，客户端并不只提取最新版本的文件快照，而是把代码仓库完整克隆下来，包括完整的历史记录。如果任何一处协同工作所使用的服务器发生故障，事后都可以用其余任何一个本地仓库进行恢复。

2.2 Git 的初识与安装

Git 是分布式管理系统中的一种，也是目前全球最流行的版本管理系统，最初版本由 Linus 研发，用于流行全球的开源操作系统 Linux 的版本管理。目前全球最有名的软件托管网站 Github、Gitee 都是依托 Git 实现的。

2.2.1 Git 的安装

进入官网https://git-scm.com/downloads，下载 Git 的 Windows 版本，双击安装，具体安装流程参考代码目录/doc/Git 安装.pdf。

2.2.2 Git 快照流

在数据处理方面，大部分系统以文件变更列表的方式存储信息，如图 2-5 所示，存储每个文件与初始版本的差异。这类系统（如 CVS 等等）将它们存储的信息看作是一组基本文件与每个文件随时间逐步累积的差异，通常称之为基于差异的版本控制。

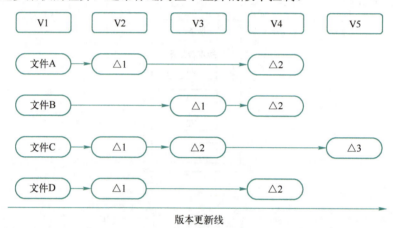

图 2-5　存储文件与初始版本的差异

Git 与上述处理数据的方式不同。Git 把数据看作是对小型文件系统的一系列快照。在 Git 中，每当用户提交更新或保存项目状态时，它基本上就会对当时的全部文件创建一个快照并保存这个快照的索引。如果文件没有修改，Git 不再重新存储该文件，而是只保留一个链接指向之前存储的文件。如图 2-6 所示，在 V2 版本中，A1 是文件 A 产生变化后的快照，在 V3 版本中，A1 虚线框代表此次快照中，文件 A 未发生变化，即产生链接指向 V2。

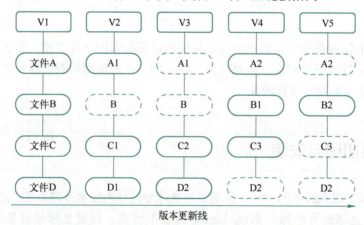

图 2-6　存储快照

2.2.3 Git 的三个区

Git 本地管理中有三个重要的区域,如图 2-7 所示,分别是工作目录、暂存区域和本地仓库。

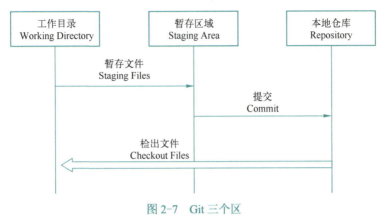

图 2-7 Git 三个区

1)工作目录:用于存放项目的某个版本独立提取出来的代码内容。

2)暂存区域:用于临时存放项目的改动信息。暂存区是一个文件,保存了下次将要提交的文件列表信息,一般在 Git 仓库目录中。

3)本地仓库:用来保存项目的元数据和对象数据库。这是 Git 中最重要的部分,从其他计算机克隆仓库时,复制的就是这里的数据。这里有项目的所有版本的数据。其中,HEAD 指向仓库最新的版本。

4)基本的 Git 工作流程如下。

① 在工作区中修改文件。

② 将你想要下次提交的更改选择性地暂存,这样只会将更改的部分添加到暂存区。

③ 提交更新,找到暂存区的文件,将快照永久性存储到 Git 目录。

对应上面三个过程文件有三种状态:已修改(modified)、已暂存(staged)和已提交(committed)。

① 已修改:自上次检出后,做了修改但还没有放到暂存区域。

② 已暂存:文件已修改并放入暂存区,还未提交,属于已暂存。

③ 已提交:Git 目录中保存着该特定版本的文件。

2.2.4 Git 基本配置

Git 一共有 3 个配置文件:

① 仓库级的配置文件:仓库的".git/.gitconfig",该配置文件只对所在的仓库有效。

② 全局配置文件:Windows 系统的"C:\Users\<用户名>\.gitconfig"。

③ 系统级的配置文件:在 Git 的安装目录下(Mac 系统下安装目录在/usr/local/git)的"etc"文件夹中的"gitconfig"。

一般的配置会使用全局配置文件。

- 查看配置信息

```
1. # --local：仓库级，--global：全局级，--system：系统级
2. $ git config [--local | --global | --system] -l
```

- 设置账户

```
1. git config --global user.name "xxxx"
2. git config --global user.email "xxxx@xxx"
```

2.3 实践 Git 基本功能

Git 最基本的功能包括对新项目进行初始化，查看当前目录的状态，可以增加新的文件到版本管理系统以及查看版本提交的日志。

2.3.1 初始化 Git 仓库

在"\path\to\works\"目录下，新建"git_study"目录，打开 Git Bash 窗口，通过 cd 命令进入 git_study 目录，然后输入"git init"命令，代码如下。

```
 1. taolee@DESKTOP-FVICLRQ MINGW64 ~/works
 2. $ cd ..
 3.
 4. taolee@DESKTOP-FVICLRQ MINGW64 ~
 5. $ cd works/git_study/
 6.
 7. taolee@DESKTOP-FVICLRQ MINGW64 ~/works/git_study
 8. $ git init
 9. Initialized empty Git repository in C:/Users/taolee/works/git_study/.git/
10.
11. taolee@DESKTOP-FVICLRQ MINGW64 ~/works/git_study (master)
12. $
```

如图 2-8 所示，在 git_study 目录下生成了".git"目录，此目录是隐藏目录，存放了 Git 版本管理所需的所有元数据和信息数据。如果看不到".git"目录，可以按照图 2-8 中的①~③步骤操作。

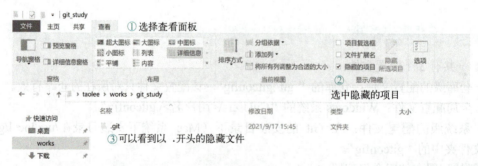

图 2-8　Git 初始化目录

2.3.2 查看当前状态

在 Git Bash 窗口中输入 git status 命令，如下所示。

```
1.  taolee@DESKTOP-FVICLRQ MINGW64 ~/works/git_study (master)
2.  $ git status
3.  On branch master
4.
5.  No commits yet
6.
7.  nothing to commit (create/copy files and use "git add" to track)
```

第 3 行，"On branch master"表示当前位于 master 分支，这是默认分支。

第 5 行，"No commits yet"表示当前还没有内容提交到本地仓库。

第 7 行，表示当前目录干净，没有需要提交的信息，可以通过创建或者复制文件，然后使用"git add"命令跟踪文件。

2.3.3 增加文件

在 git_study 目录增加说明文件 Readme.md，输入如下内容。

```
1.  # Git Study
2.
3.  ## 这是 git 的学习目录
4.
5.  这是第一次实践 git
```

然后在 Git Bash 窗口里输入命令"git status"，代码如下。

```
1.  taolee@DESKTOP-FVICLRQ MINGW64 ~/works/git_study (master)
2.  $ git status
3.  On branch master
4.
5.  No commits yet
6.
7.  Untracked files:
8.    (use "git add <file>..." to include in what will be committed)
9.        Readme.md
10.
11. nothing added to commit but untracked files present (use "git add" to track)
```

第 3 行，表示当前位于 master 分支。

第 5 行，表示当前还没有可提交内容。

第 7～9 行，表示当前有新文件，Readme.md 未加入跟踪，使用"git add 文件名"将新文件加入 Git 版本管理。"未跟踪"文件，是指那些新添加的并且未被加入到暂存区域或提交的文件。Git 只处理加入版本管理的文件，所以需要通过 git add 命令将新文件加入跟踪。

第 11 行，表示当前没有需要提交到仓库的内容，但有未跟踪的文件，使用 git add 命令跟踪文件。

使用"git add [files]"将文件加入跟踪，即加入暂存区，代码如下。

```
1.  taolee@DESKTOP-FVICLRQ MINGW64 ~/works/git_study (master)
2.  $ git add Readme.md
3.
4.  taolee@DESKTOP-FVICLRQ MINGW64 ~/works/git_study (master)
5.  $ git status
6.  On branch master
7.
8.  No commits yet
9.
10. Changes to be committed:
11.   (use "git rm --cached <file>..." to unstage)
12.         new file:   Readme.md
```

第 2 行，将 Readme.md 文件加入跟踪。

第 5 行，再次输入 git status 命令。

第 6 行，表示当前位于 master 分支。

第 8 行，表示还没有提交过。

第 10～12 行，表示暂存区有改动，需要提交到本地仓库，如果此时不想将文件加入暂存区，可以使用"git rm --cached <file>..."将暂存区文件删除。新加入暂存区的文件是 Readme.md。

使用"git rm --cached Readme.md"撤销上一步操作，代码如下。

```
1.  taolee@DESKTOP-FVICLRQ MINGW64 ~/works/git_study (master)
2.  $ git rm --cached Readme.md
3.  rm 'Readme.md'
4.
5.  taolee@DESKTOP-FVICLRQ MINGW64 ~/works/git_study (master)
6.  $ git status
7.  On branch master
8.
9.  No commits yet
10.
11. Untracked files:
12.   (use "git add <file>..." to include in what will be committed)
13.         Readme.md
14.
15. nothing added to commit but untracked files present (use "git add" to track)
```

第 2 行，撤销 Readme.md 加入暂存区的操作。

第 3 行，表示删除 Readme.md 文件。

第 6 行，再次输入 git status 查看当前状态。

第 7～15 行，又回到新增加 Readme.md 文件，而未加入暂存区的状态了。

再次加入暂存区，并使用"git commit -m '注释'"提交修改到仓库，代码如下。

```
1.  taolee@DESKTOP-FVICLRQ MINGW64 ~/works/git_study (master)
2.  $ git add Readme.md
3.
4.  taolee@DESKTOP-FVICLRQ MINGW64 ~/works/git_study (master)
```

```
5.    $ git commit -m "增加 Readme.md 文件"
6.   [master (root-commit) 727691d] 增加 Readme.md 文件
7.    1 file changed, 5 insertions(+)
8.    create mode 100644 Readme.md
9.
10.  taolee@DESKTOP-FVICLRQ MINGW64 ~/works/git_study (master)
11.  $ git status
12.  On branch master
13.  nothing to commit, working tree clean
```

第 5 行，将修改提交到本地仓库。

第 6～8 行，表示 1 个文件改动，插入了 5 行新内容。

第 11～13 行，表示当前工作目录干净，不用提交。

打开 Readme.md 文件，加入新内容第 6 行。

```
1.   # Git Study
2.
3.   ## 这是 git 的学习目录
4.
5.   这是第一次实践 git
6.   增加新的一行
```

再次运行 git status，代码如下。

```
1.   taolee@DESKTOP-FVICLRQ MINGW64 ~/works/git_study (master)
2.   $ git status
3.   On branch master
4.   Changes not staged for commit:
5.     (use "git add <file>..." to update what will be committed)
6.     (use "git restore <file>..." to discard changes in working directory)
7.           modified:   Readme.md
8.
9.   no changes added to commit (use "git add" and/or "git commit -a")
```

第 5 行，表示使用 "git add <file>..." 将修改的内容从工作目录更新到暂存区。

第 6 行，表示可以使用 "git restore <file>..." 放弃修改。

第 7 行，列出改动的文件。

第 9 行，表示可以分步骤，先通过 "git add" 将改动加入暂存区，然后再通过 "git commit" 提交修改。或者一步到位，使用 git commit -am 自动包含将改动加入暂存区，然后提交到仓库。

使用 "git commit -am '注释'"，代码如下。

```
1.   taolee@DESKTOP-FVICLRQ MINGW64 ~/works/git_study (master)
2.   $ git commit -am "Readme 增加了一行内容"
3.   [master c016103] Readme 增加了一行内容
4.    1 file changed, 2 insertions(+), 1 deletion(-)
5.
6.   taolee@DESKTOP-FVICLRQ MINGW64 ~/works/git_study (master)
7.   $ git status
8.   On branch master
9.   nothing to commit, working tree clean
```

第 2 行，使用 commit 的 -am 选项，一步到位，将修改提交到仓库。

2.3.4 查看 log

在 Git Bash 窗口中输入 git log，如下。

```
1.   taolee@DESKTOP-FVICLRQ MINGW64 ~/works/git_study (master)
2.   $ git log
3.   commit c016103349cceeb20199f8e95c497755dcfb1638 (HEAD -> master)
4.   Author: taolee <taolee01@126.com>
5.   Date:   Fri Sep 17 16:18:55 2021 +0800
6.
7.       Readme 增加了一行内容
8.
9.   commit 727691dd8c2b1910fcade985ca6477a0bde0eb19
10.  Author: taolee <taolee01@126.com>
11.  Date:   Fri Sep 17 16:15:56 2021 +0800
12.
13.      增加 Readme.md 文件
```

按提交的逆序打印提交的版本，包含了 commit id、作者、日期及注释等信息。

2.4 Git 版本回退

在开发的过程当中，经常会由于某种原因需要回退到以前的某个版本，Git 支持该功能，可以很方便地进行版本回退。

2.4.1 commit id

Git 会给每一次提交标识一个唯一的 commit id。这样就允许我们通过 commit id 来定位到任何一次提交。一个完整的 commit id 包含 40 个字符。

例如，c016103349cceeb20199f8e95c497755dcfb1638。

Git 提供了一些速记符号，可以高效地定位历史提交，速记符如表 2-1 所示，"HEAD"指向当前分支最新入库的版本。也可以直接使用 commit id 的一部分来定位一次提交，通常情况下只需要提供一个 commit id 中的 4~5 个字符，Git 就能识别出对应的提交了，可以通过 git reflog 查看。

表 2-1 速记符

速记符	含义
HEAD	最后一次提交
HEAD^	前一次提交
HEAD^^	之前第二次提交
HEAD~1	前一次提交
HEAD~2	之前第二次提交

注：HEAD 表示最新提交的快照，而 HEAD~ 表示 HEAD 的上一个快照，如果表示上 5 个快照，则可以用 HEAD~5。

2.4.2 reset 命令

若想对版本进行回退，可以通过 reset 命令，如表 2-2 所示。

表 2-2 reset 回退

命令	说明
git reset --soft commit-id	只改变仓库提交点，暂存区和工作目录的内容都不改变
git reset --mixed commit-id	改变仓库提交点，同时改变暂存区的内容（默认）
git reset --hard commit-id	暂存区、工作目录的内容都会被修改到与提交点完全一致的状态
git reset --hard HEAD	让工作目录回到上次提交时的状态

以上命令如果没有指定 commit id，则默认为当前 HEAD。

如果此时，想让工作目录与仓库最新版本一样，可以通过如下代码实现。

```
1.  taolee@DESKTOP-FVICLRQ MINGW64 ~/works/git_study (master)
2.  $ echo "hello world" >> Readme.md
3.
4.  taolee@DESKTOP-FVICLRQ MINGW64 ~/works/git_study (master)
5.  $ cat Readme.md
6.  # Git Study
7.
8.  ## 这是 git 的学习目录
9.
10. 这是第一次实践 git
11. 增加新的一行 hello world
12. taolee@DESKTOP-FVICLRQ MINGW64 ~/works/git_study (master)
13. $ git status
14. On branch master
15. Changes not staged for commit:
16.   (use "git add <file>..." to update what will be committed)
17.   (use "git restore <file>..." to discard changes in working directory)
18.         modified:   Readme.md
19.
20. no changes added to commit (use "git add" and/or "git commit -a")
21.
22. taolee@DESKTOP-FVICLRQ MINGW64 ~/works/git_study (master)
23. $ git reset --hard
24. HEAD is now at c016103 Readme 增加了一行内容
25.
26. taolee@DESKTOP-FVICLRQ MINGW64 ~/works/git_study (master)
27. $ git status
28. On branch master
29. nothing to commit, working tree clean
```

第 2 行，通过 echo 命令，向 Readme.md 文件末尾增加一行内容。

第 5～11 行，通过 cat 命令打印文件内容。

第 13～21 行，通过打印当前工作目录状态，可以看到工作目录文件修改了。

第 23 行，通过 git reset --hard 命令将仓库最新内容覆盖暂存区和工作目录。

第 27~29 行，打印工作目录状态，可以看到工作目录修改的内容丢失，回到仓库的版本。
reset 命令不仅可以回退版本，也可以回退指定文件。

命令格式为：git reset commit-id <file-path>

该命令不会移动 HEAD 指向，而是直接将指定快照的指定文件回退到暂存区域。

2.5　Git 内容对比

随着新功能的不断加入，有时需要查看新增加的功能涉及哪些文件，以及具体文件里的修改内容，这时候需要一种可以记录差异，并可以查看新增内容与旧版本差异的功能，Git 可以提供这些功能。

2.5.1　对比工作目录与暂存区

将 git_study 目录里的 Readme.md 进行修改，原先内容如下。

```
1.   # Git Study
2.   
3.   ## 这是 git 的学习目录
4.   
5.   这是第一次实践 git
6.   增加新的一行
```

修改后如下。

```
1.   # Git Study
2.   
3.   ## 这是 git 的学习目录
4.   
5.   这是第一次实践 git 修改此行
6.   增加新的一行
7.   继续增加第二行
```

可以看到修改了第 5 行，同时增加了第 7 行内容。

在 Git Bash 窗口里输入命令 "git diff"，内容如下。

```
1.   taolee@DESKTOP-FVICLRQ MINGW64 ~/works/git_study (master)
2.   $ git diff
3.   diff --git a/Readme.md b/Readme.md
4.   index 1eaa608..00a01d8 100644
5.   --- a/Readme.md
6.   +++ b/Readme.md
7.   @@ -2,5 +2,6 @@
8.   
9.    ## 这是 git 的学习目录
10.   
11.   -这是第一次实践 git
12.   -增加新的一行
```

```
13.  \ No newline at end of file
14.  +这是第一次实践 git 修改此行
15.  +增加新的一行
16.  +继续增加第二行
17.  \ No newline at end of file
```

第 2 行，输入对比命令。

第 3 行，表示对比暂存区 a/Readme.md 与工作目录 b/Readme.md。

第 4 行，对比文件的 ID，左边的是暂存区的，右边是工作目录，100644 是文件的权限和类型。

第 5 行，表示暂存区的旧文件。

第 6 行，表示工作目录中修改的文件。

第 7 行，表示以 @@ 开头和结束，中间的"-"表示旧文件，"+"表示新文件，后边的数字表示"开始行号，连续行数"。

第 11~12 行，-表示新文件中这两行已经删除。

第 13 行，表示此处文件没有换行符，只是 Git 为了显示友好，增加了说明文字。

第 14~16 行，表示新文件新增内容，Git 将两个文件合并显示，以旧文件为参考，旧文件通过增删等变成新文件。

第 17 行，表示该文件结尾没有换行符，Git 为了显示友好，增加了说明文字。

如果结果中出现^M，这是由于不同操作系统对换行符的处理差异导致的，Windows 用 CRLF 来定义换行，Linux 用 LF。CR 全称是 Carriage Return，或者表示为\r，意思是回车。LF 全称是 Line Feed，它才是真正意义上的换行符。可以配置 Git 的全局参数，示例如下。

```
1.  taolee@DESKTOP-FVICLRQ MINGW64 ~/works/git_study (master)
2.  $ git config --global core.whitespace cr-at-eol
```

2.5.2 对比仓库不同版本

通过 git log 命令查看仓库提交历史，代码如下。

```
1.  taolee@DESKTOP-FVICLRQ MINGW64 ~/works/git_study (master)
2.  $ git log
3.  commit c016103349cceeb20199f8e95c497755dcfb1638 (HEAD -> master)
4.  Author: taolee <taolee01@126.com>
5.  Date:   Fri Sep 17 16:18:55 2021 +0800
6.
7.      Readme 增加了一行内容
8.
9.  commit 727691dd8c2b1910fcade985ca6477a0bde0eb19
10. Author: taolee <taolee01@126.com>
11. Date:   Fri Sep 17 16:15:56 2021 +0800
12.
13.     增加 Readme.md 文件
```

总共提交了两次，通过查看第 3 行和第 9 行的前 5 位数字，可以定位这两个版本，c0161，72769。

通过"git diff commit-id1 commit-id2"可以对比仓库中两个版本的差异，代码如下。

```
1.   taolee@DESKTOP-FVICLRQ MINGW64 ~/works/git_study (master)
2.   $ git diff 72769 c01610
3.   diff --git a/Readme.md b/Readme.md
4.   index 36f4286..1eaa608 100644
5.   --- a/Readme.md
6.   +++ b/Readme.md
7.   @@ -2,4 +2,5 @@
8.
9.    ## 这是git 的学习目录
10.
11.  -这是第一次实践git
12.  \ No newline at end of file
13.  +这是第一次实践git
14.  +增加新的一行
15.  \ No newline at end of file
```

可以看到仓库里两次提交版本之间的文件 Readme.md 的差异。

2.5.3 对比工作目录与仓库

通过 git diff commit id 可以查看指定仓库版本与当前目录差异。

```
1.   taolee@DESKTOP-FVICLRQ MINGW64 ~/works/git_study (master)
2.   $ git diff c0161
3.   diff --git a/Readme.md b/Readme.md
4.   index 1eaa608..00a01d8 100644
5.   --- a/Readme.md
6.   +++ b/Readme.md
7.   @@ -2,5 +2,6 @@
8.
9.    ## 这是git 的学习目录
10.
11.  -这是第一次实践git
12.  -增加新的一行
13.  \ No newline at end of file
14.  +这是第一次实践git 修改此行
15.  +增加新的一行
16.  +继续增加第二行
17.  \ No newline at end of file
```

可以看到当前目录内容比仓库 c0161 版本新。

2.5.4 对比暂存区与仓库

通过 git diff --cached [commit id]可以查看暂存区与仓库版本区别，如果 commit id 省略，则对比暂存区与仓库最新版本区别。

```
1.   taolee@DESKTOP-FVICLRQ MINGW64 ~/works/git_study (master)
```

```
2.  $ git add Readme.md
3.
4.  taolee@DESKTOP-FVICLRQ MINGW64 ~/works/git_study (master)
5.  $ git diff --cached
6.  diff --git a/Readme.md b/Readme.md
7.  index 1eaa608..00a01d8 100644
8.  --- a/Readme.md
9.  +++ b/Readme.md
10. @@ -2,5 +2,6 @@
11.
12.   ## 这是git的学习目录
13.
14. -这是第一次实践git
15. -增加新的一行
16. \ No newline at end of file
17. +这是第一次实践git 修改此行
18. +增加新的一行
19. +继续增加第二行
20. \ No newline at end of file
```

第 2 行，将当前工作目录内容提交到暂存区。

第 5 行，比较暂存区与仓库最新提交版本的差异。

第 6～20 行，表示暂存区对比仓库版本有改动。

最后，通过 git commit -m "Readme 增加第二行" 提交暂存区内容到仓库。

2.6 Git 分支与查看

项目有时需要从开发主线上分离开来，以免影响开发主线，比如定制新版本，修改某个 bug，测试某种特性功能等。需要通过建立新的分支，实现多分支管理。

2.6.1 分支创建

分支创建可以通过命令"git branch 分支名"实现。

```
1.  taolee@DESKTOP-FVICLRQ MINGW64 ~/works/git_study (master)
2.  $ git branch v1
3.
4.  taolee@DESKTOP-FVICLRQ MINGW64 ~/works/git_study (master)
5.  $ git branch
6.  * master
7.    v1
```

第 2 行，增加了新的分支 v1。

第 5～7 行，查看分支，有 master 分支与 v1 分支，其中*表示当前所在分支。

```
1.  taolee@DESKTOP-FVICLRQ MINGW64 ~/works/git_study (master)
```

```
2.   $ git log --decorate
3.   commit ba4c9ddbe76b03992c6c3c4949bb5ccb99a47a57 (HEAD -> master, v1)
4.   Author: taolee <taolee01@126.com>
5.   Date:   Fri Sep 17 17:11:06 2021 +0800
6.
7.       Readme 增加第二行
8.
9.   commit c016103349cceeb20199f8e95c497755dcfb1638
10.  Author: taolee <taolee01@126.com>
11.  Date:   Fri Sep 17 16:18:55 2021 +0800
12.
13.      Readme 增加了一行内容
14.
15.  commit 727691dd8c2b1910fcade985ca6477a0bde0eb19
16.  Author: taolee <taolee01@126.com>
17.  Date:   Fri Sep 17 16:15:56 2021 +0800
18.
19.      增加 Readme.md 文件
```

通过 git log --decorate 命令，可以看到仓库当前版本信息。

第 3 行，HEAD 指向 master 分支，由于 v1 是基于 master 分支创建的，所以此时刻它们位置一致。

在当前工作目录增加新文件，并提交到仓库，代码如下。

```
1.  taolee@DESKTOP-FVICLRQ MINGW64 ~/works/git_study (master)
2.  $ touch 1.c
3.
4.  taolee@DESKTOP-FVICLRQ MINGW64 ~/works/git_study (master)
5.  $ git add .
6.
7.  taolee@DESKTOP-FVICLRQ MINGW64 ~/works/git_study (master)
8.  $ git commit -m "增加 1.c 文件"
9.  [master 3e34ce1] 增加 1.c 文件
10.   1 file changed, 0 insertions(+), 0 deletions(-)
11.   create mode 100644 1.c
12.
13. taolee@DESKTOP-FVICLRQ MINGW64 ~/works/git_study (master)
14. $ git log --decorate --oneline --graph --all
15. * 3e34ce1 (HEAD -> master) 增加 1.c 文件
16. * ba4c9dd (v1) Readme 增加第二行
17. * c016103 Readme 增加了一行内容
18. * 727691d 增加 Readme.md 文件
```

第 2 行，通过 touch 命令在当前工作目录创建新的文件 1.c。

第 4~11 行，将新文件提交到仓库。

第 14 行，以简化格式打印仓库情况。

第 15~16 行，表示 master 分支比 v1 分支多了新的文件，同时可以看到 HEAD 指针指向当前分支。

2.6.2 分支切换

可以通过"git checkout 分支名"切换指定分支。

```
1.   taolee@DESKTOP-FVICLRQ MINGW64 ~/works/git_study (master)
2.   $ git checkout v1
3.   Switched to branch 'v1'
4.
5.   taolee@DESKTOP-FVICLRQ MINGW64 ~/works/git_study (v1)
6.   $ git branch
7.     master
8.   * v1
9.
10.  taolee@DESKTOP-FVICLRQ MINGW64 ~/works/git_study (v1)
11.  $ git log --oneline --decorate --graph --all
12.  * 3e34ce1 (master) 增加 1.c 文件
13.  * ba4c9dd (HEAD -> v1) Readme 增加第二行
14.  * c016103 Readme 增加了一行内容
15.  * 727691d 增加 Readme.md 文件
```

第 2 行,通过"git checkout v1"命令切换到 v1 分支。

第 6 行,通过 git branch 查看分支,可以看到*位于 v1。

第 11 行,以简化格式打印日志。

第 13 行,HEAD 指向了 v1,表示当前位于 v1 分支。

在 v1 分支做一些修改,代码如下。

```
1.   taolee@DESKTOP-FVICLRQ MINGW64 ~/works/git_study (v1)
2.   $ touch 2.c
3.
4.   taolee@DESKTOP-FVICLRQ MINGW64 ~/works/git_study (v1)
5.   $ git add .
6.
7.   taolee@DESKTOP-FVICLRQ MINGW64 ~/works/git_study (v1)
8.   $ git commit -m "v1 分支新增 2.c 文件"
9.   [v1 fc35525] v1 分支新增 2.c 文件
10.   1 file changed, 0 insertions(+), 0 deletions(-)
11.   create mode 100644 2.c
12.
13.  taolee@DESKTOP-FVICLRQ MINGW64 ~/works/git_study (v1)
14.  $ git log --oneline --decorate --graph --all
15.  * fc35525 (HEAD -> v1) v1 分支新增 2.c 文件
16.  | * 3e34ce1 (master) 增加 1.c 文件
17.  |/
18.  * ba4c9dd Readme 增加第二行
19.  * c016103 Readme 增加了一行内容
20.  * 727691d 增加 Readme.md 文件
```

第 2 行,增加文件 2.c。

第 4~11 行,将增加内容同步到仓库。

第 14 行，打印日志。

第 15～20 行，可以看到 v1 分支与 master 分支向各自方向发展。

2.6.3 分支合并

通过"git merge 分支名"可以将指定分支合并到当前分支。

```
1.  taolee@DESKTOP-FVICLRQ MINGW64 ~/works/git_study (v1)
2.  $ git checkout master
3.  Switched to branch 'master'
4.
5.  taolee@DESKTOP-FVICLRQ MINGW64 ~/works/git_study (master)
6.  $ git merge v1
7.  Merge made by the 'recursive' strategy.
8.   2.c | 0
9.   1 file changed, 0 insertions(+), 0 deletions(-)
10.  create mode 100644 2.c
11.
12. taolee@DESKTOP-FVICLRQ MINGW64 ~/works/git_study (master)
13. $ git log --oneline --decorate --graph --all
14. *   8260e8c (HEAD -> master) Merge branch 'v1'
15. |\
16. | * fc35525 (v1) v1 分支新增 2.c 文件
17. * | 3e34ce1 增加 1.c 文件
18. |/
19. * ba4c9dd Readme 增加第二行
20. * c016103 Readme 增加了一行内容
21. * 727691d 增加 Readme.md 文件
```

第 3 行，切换回 mater 分支。

第 6 行，将 v1 分支合并到 master 分支。

第 14～21 行，可以看到 master 分支合并了 v1 分支的功能。

在合并分支的过程当中，可能存在冲突，可以先解决冲突，再合并。

2.7 Git 标签管理

Git 可以给仓库历史中的某一个提交版本打上标签，以此来方便管理各版本。

2.7.1 查看标签

通过 git tag 命令可以查看本地仓库的标签列表。

2.7.2 创建标签

Git 支持两种标签：轻量标签（lightweight）与注释标签（annotated）。轻量标签是某个特定提交的引用。注释标签是存储在 Git 数据库中的一个完整对象，它们是可以校验的，其中包含打标签者的名字、电子邮件地址、日期时间，此外还有一个标签信息，并且可以使用 GNU Privacy Guard（GPG）签名并验证。建议一般情况使用注释标签。

```
1.   taolee@DESKTOP-FVICLRQ MINGW64 ~/works/git_study (master)
2.   $ git tag -a v2.0 -m "这是注释标签，v2.0 版本发布"
3.
4.   taolee@DESKTOP-FVICLRQ MINGW64 ~/works/git_study (master)
5.   $ git tag
6.   v2.0
```

第 2 行，对当前 master 分支最新版本打标签。-m 选项指定了一条将会存储在标签中的信息。如果没有为附注的标签指定一条信息，Git 会启动编辑器要求你输入信息。

第 5 行，查看当前仓库的标签列表。

通过使用 git show 命令可以看到标签信息和与之对应的提交信息。

```
1.   taolee@DESKTOP-FVICLRQ MINGW64 ~/works/git_study (master)
2.   $ git show v2.0
3.   tag v2.0
4.   Tagger: taolee <taolee01@126.com>
5.   Date:   Fri Sep 17 17:55:30 2021 +0800
6.
7.   这是注释标签，v2.0 版本发布
8.
9.   commit 8260e8c86c84f9084ac806a5e30c16cb17050310 (HEAD -> master, tag: v2.0)
10.  Merge: 3e34ce1 fc35525
11.  Author: taolee <taolee01@126.com>
12.  Date:   Fri Sep 17 17:44:02 2021 +0800
13.
14.      Merge branch 'v1'
```

结果中显示了打标签者的信息、打标签的日期时间、附注信息，然后显示具体的提交信息。

2.7.3 指定版本打标签

可以通过 "git tag -a 标签名 commit-id -m 标签注释" 给指定版本打标签。

```
1.   taolee@DESKTOP-FVICLRQ MINGW64 ~/works/git_study (master)
2.   $ git log --pretty=oneline
3.   8260e8c86c84f9084ac806a5e30c16cb17050310 (HEAD -> master, tag: v2.0) Merge branch 'v1'
4.   fc35525ef6e29cfff0f77a888c4df6f5a6383491 (v1) v1 分支新增 2.c 文件
5.   3e34ce183a64779309cebb8c315b343b9a16d838 增加 1.c 文件
6.   ba4c9ddbe76b03992c6c3c4949bb5ccb99a47a57 Readme 增加第二行
7.   c016103349cceeb20199f8e95c497755dcfb1638 Readme 增加了一行内容
8.   727691dd8c2b1910fcade985ca6477a0bde0eb19 增加 Readme.md 文件
9.
10.  taolee@DESKTOP-FVICLRQ MINGW64 ~/works/git_study (master)
11.  $ git tag -a v1.0 fc355 -m "发布 v1.0 版本"
12.
```

```
13. taolee@DESKTOP-FVICLRQ MINGW64 ~/works/git_study (master)
14. $ git log --pretty=oneline
15. 8260e8c86c84f9084ac806a5e30c16cb17050310 (HEAD -> master, tag: v2.0) Merge branch 'v1'
16. fc35525ef6e29cfff0f77a888c4df6f5a6383491 (tag: v1.0, v1) v1 分支新增 2.c 文件
17. 3e34ce183a64779309cebb8c315b343b9a16d838 增加 1.c 文件
18. ba4c9ddbe76b03992c6c3c4949bb5ccb99a47a57 Readme 增加第二行
19. c016103349cceeb20199f8e95c497755dcfb1638 Readme 增加了一行内容
20. 727691dd8c2b1910fcade985ca6477a0bde0eb19 增加 Readme.md 文件
```

2.7.4 检出标签

通过"**git checkout 标签名**",可以从仓库检出指定标签的版本。

```
1. taolee@DESKTOP-FVICLRQ MINGW64 ~/works/git_study (master)
2. $ git checkout v1.0
3. Note: switching to 'v1.0'.
4.
5. You are in 'detached HEAD' state. You can look around, make experimental
6. changes and commit them, and you can discard any commits you make in this
7. state without impacting any branches by switching back to a branch.
8.
9. If you want to create a new branch to retain commits you create, you may
10. do so (now or later) by using -c with the switch command. Example:
11.
12.   git switch -c <new-branch-name>
13.
14. Or undo this operation with:
15.
16.   git switch -
17.
18. Turn off this advice by setting config variable advice.detachedHead to false
19.
20. HEAD is now at fc35525 v1 分支新增 2.c 文件
21.
22. taolee@DESKTOP-FVICLRQ MINGW64 ~/works/git_study ((v1.0))
23. $ git log --oneline --decorate --graph --all
24. *   8260e8c (tag: v2.0, master) Merge branch 'v1'
25. |\
26. | * fc35525 (HEAD, tag: v1.0, v1) v1 分支新增 2.c 文件
27. * | 3e34ce1 增加 1.c 文件
28. |/
29. * ba4c9dd Readme 增加第二行
30. * c016103 Readme 增加了一行内容
31. * 727691d 增加 Readme.md 文件
32.
33. taolee@DESKTOP-FVICLRQ MINGW64 ~/works/git_study ((v1.0))
34. $ git branch
```

```
35.    * (HEAD detached at v1.0)
36.      master
37.      v1
```

第 2~21 行，检出 v1.0 版本。输出的内容，此时我们处于匿名分支，即当前 HEAD 指针处于游离状态，一旦切换到其他分支，对当前版本的修改，就会丢弃。

第 26 行，可以看到 HEAD 指向了 tag:v1.0。

2.7.5 删除标签

通过"git tag -d 标签名"，可以删除指定标签。

```
1.  taolee@DESKTOP-FVICLRQ MINGW64 ~/works/git_study ((v1.0))
2.  $ git checkout master
3.  Previous HEAD position was fc35525 v1 分支新增 2.c 文件
4.  Switched to branch 'master'
5.
6.  taolee@DESKTOP-FVICLRQ MINGW64 ~/works/git_study (master)
7.  $ git tag -d v1.0
8.  Deleted tag 'v1.0' (was 0a84aee)
9.
10. taolee@DESKTOP-FVICLRQ MINGW64 ~/works/git_study (master)
11. $ git tag
12. v2.0
```

第 2 行，切换回 master 分支。

第 7 行，删除标签 v1.0。

第 11 行，查看当前仓库标签列表。可以看到 v1.0 标签已经删除。

2.8 将工程加入 Git 版本控制

使用本章所学的 Git 版本管理知识，创建初始提交，并将任务 1 的工程加入到 Git 版本管理系统。

2.8.1 初始化工作目录

打开 Git Bash，使用 cd 命令，切换目录到工程目录"/path/to/SmartAgriculture"，输入 git init，初始化版本控制，具体代码如下。

```
1.  taolee@DESKTOP-0HJVM6A MINGW64 ~/works/HarmonyOS/Books/SmartAgriculture
2.  $ git init
3.  Initialized empty Git repository in C:/Users/taolee/works/HarmonyOS/Books/SmartAgriculture/.git/
4.
5.  taolee@DESKTOP-0HJVM6A MINGW64 ~/works/HarmonyOS/Books/SmartAgriculture (master)
6.  $ ls
7.  LICENSE   README.md
```

```
 8.
 9. taolee@DESKTOP-OHJVM6A MINGW64 ~/works/HarmonyOS/Books/SmartAgriculture
(master)
10. $ git status
11. On branch master
12.
13. No commits yet
14.
15. Untracked files:
16.   (use "git add <file>..." to include in what will be committed)
17.         LICENSE
18.         README.md
19.
20. nothing added to commit but untracked files present(use "git add"to track)
21.
22. taolee@DESKTOP-OHJVM6A MINGW64 ~/works/HarmonyOS/Books/SmartAgriculture
(master)
23. $ git add .
24. taolee@DESKTOP-OHJVM6A MINGW64 ~/works/HarmonyOS/Books/SmartAgriculture
(master)
25. $ git commit -m "Initial commit"
26. [master (root-commit) c877574] Initial commit
27.  2 files changed, 272 insertions(+)
28.  create mode 100644 LICENSE
29.  create mode 100644 README.md
30.
31. taolee@DESKTOP-OHJVM6A MINGW64 ~/works/HarmonyOS/Books/SmartAgriculture
(master)
32. $ git status
33. On branch master
34. nothing to commit, working tree clean
35.
36. taolee@DESKTOP-OHJVM6A MINGW64 ~/works/HarmonyOS/Books/SmartAgriculture
(master)
37. $ git log --pretty=oneline
38. c8775743058113b200cb2e20a370e25a9056b365 (HEAD -> master) Initial commit
```

第 2 行，通过 git init 命令初始化当前工作目录，并将当前目录加入到 Git 版本管理系统。

第 6 行，通过 ls 命令可以查看当前目录下存在的文件。在当前目录下新建了 README.md 和 LICENSE 文件，其中 README.md 用于对当前项目的说明，一般采用 Markdown 格式，LICENSE 文件用于对当前项目的协议进行说明。

第 10 行，通过"git status"命令查看当前项目的状态，可以看到新文件 README.md 和 LICENSE 未加入到版本系统。

第 23 行，通过"git add ."命令，将当前项目的变动文件加入到暂存状态。

第 25 行，通过"git commit -m "Initial commit""命令，将当前项目的暂存状态提交到本地仓库。

第 32 行，通过"git status"命令查看当前项目的状态，可以看到项目目录干净。

第 37 行，通过"git log --pretty=oneline"命令查看提交的历史日志。

README.md 文件一般用于对项目进行必要的说明，如下。

```
1.   # 智慧农业
2.   
3.   本项目是配套`《HarmonyOS 应用开发与实践》`一书的代码。
4.   
5.   ## 作者信息
6.   
7.   <table border="1">
8.       <tr>
9.           <td>主编</td>
10.          <td>季云峰</td>
11.          <td>李涛</td>
12.          <td colspan="2">高云</td>
13.      </tr>
14.      <tr>
15.          <td>参编</td>
16.          <td>刘丽</td>
17.          <td>徐威</td>
18.          <td>匡亮</td>
19.          <td>平震宇</td>
20.      </tr>
21.      <tr>
22.          <td>作者单位</td>
23.          <td  colspan="4">江苏信息职业技术学院</td>
24.      </tr>
25.  </table>
26.  
27.  ## `《HarmonyOS 应用开发与实践》`书目录
28.  
29.  ```shell
30.  绪论   系统概述及设计
31.      0.1  项目背景
32.      0.2  项目方案
33.      0.3  系统部署
34.      0.4  系统功能
35.          0.4.1  Splash 欢迎界面
36.          0.4.2  系统登录界面
37.          0.4.3  大气环境监控界面
38.          0.4.4  土壤环境监控界面
39.          0.4.5  水阀控制界面
40.          0.4.6  参数设置界面
41.  任务 1   开发环境搭建和创建项目
42.  
43.  ……省略……
44.  
45.  任务 14   创建多语言
46.      14.1   多语言设计
47.  
48.  
49.  
50.  
```

```
51.     14.2    全屏显示
52.
53.
54.
55.     14.3    退出当前账号
56.         14.3.1      更新 MyApplication.java 文件
57.         14.3.2      更新 SplashAbilitySlice.java 文件
58.         14.3.3      更新 LoginAbilitySlice.java 文件
59.         14.3.4      更新 SplashAbility.java 文件
60.         14.3.5      更新 MainAbilitySlice.java 文件
61.         14.3.6      了解 Page Ability 生命周期
62.         14.3.7      编译运行
63.     14.4    提交代码到仓库
64.
65. ```
```

智慧农业的 README.md 采用了 Markdown 和 HTML 语言对项目的名称、作者、配套书籍目录进行了说明。

LICENSE 文件暂且留空，对于需要相关协议说明的，可以写入该文件。

2.8.2 提交代码到本地仓库

将任务 1 的代码复制到当前目录，如图 2-9 所示。

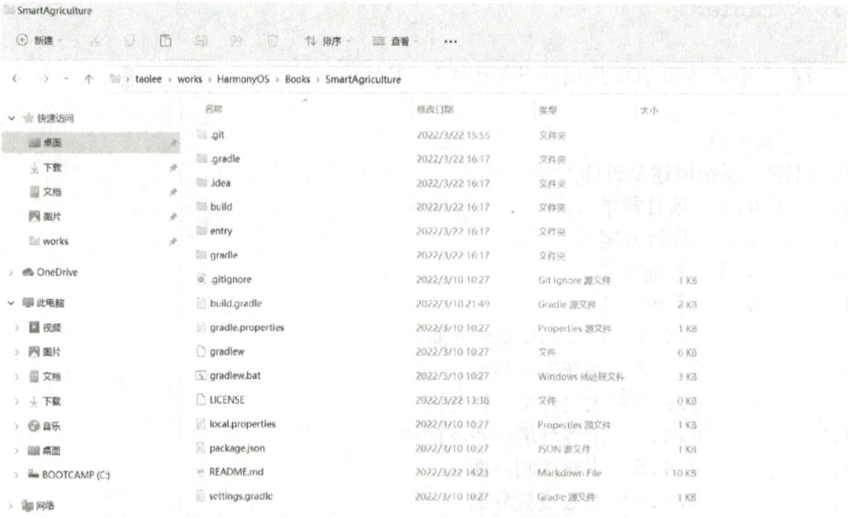

图 2-9　SmartAgriculture 目录

将任务 1 的代码加入到版本管理系统。

```
1.  taolee@DESKTOP-0HJVM6A MINGW64 ~/works/HarmonyOS/Books/SmartAgriculture (master)
2.  $ git status
3.  On branch master
4.  Untracked files:
5.    (use "git add <file>..." to include in what will be committed)
```

```
 6.         .gitignore
 7.         .idea/
 8.         build.gradle
 9.         entry/
10.         gradle.properties
11.         gradle/
12.         gradlew
13.         gradlew.bat
14.         package.json
15.         settings.gradle
16.
17. nothing added to commit but untracked files present (use "git add"to track)
18.
19. taolee@DESKTOP-0HJVM6A MINGW64 ~/works/HarmonyOS/Books/SmartAgriculture (master)
20. $ git add .
21. warning: LF will be replaced by CRLF in gradle.properties.
22. The file will have its original line endings in your working directory
23. warning: LF will be replaced by CRLF in gradle/wrapper/gradle-wrapper.properties.
24. The file will have its original line endings in your working directory
25. warning: LF will be replaced by CRLF in gradlew.
26. The file will have its original line endings in your working directory
27. warning: LF will be replaced by CRLF in gradlew.bat.
28. The file will have its original line endings in your working directory
29.
30. taolee@DESKTOP-0HJVM6A MINGW64 ~/works/HarmonyOS/Books/SmartAgriculture (master)
31. $ git commit -m "更改App图标和名称"
32. [master 4df17b7] 更改App图标和名称
33.  30 files changed, 649 insertions(+)
34.  create mode 100644 .gitignore
35.  create mode 100644 .idea/.gitignore
36.  create mode 100644 .idea/compiler.xml
37.  create mode 100644 .idea/gradle.xml
38.  create mode 100644 .idea/jarRepositories.xml
39.  create mode 100644 .idea/misc.xml
40.  create mode 100644 build.gradle
41.  create mode 100644 entry/.gitignore
42.  create mode 100644 entry/build.gradle
43.  create mode 100644 entry/proguard-rules.pro
44.  create mode 100644 entry/src/main/config.json
45.  create mode 100644 entry/src/main/java/com/example/smartagriculture/MainAbility.java
46.  create mode 100644 entry/src/main/java/com/example/smartagriculture/MyApplication.java
47.  create mode 100644 entry/src/main/java/com/example/smartagriculture/slice/MainAbilitySlice.java
48.  create mode 100644 entry/src/main/resources/base/element/string.json
```

```
49.  create mode 100644 entry/src/main/resources/base/graphic/background_ability_main.xml
50.  create mode 100644 entry/src/main/resources/base/layout/ability_main.xml
51.  create mode 100644 entry/src/main/resources/base/media/icon.png
52.  create mode 100644 entry/src/main/resources/base/media/logo.png
53.  create mode 100644 entry/src/main/resources/en/element/string.json
54.  create mode 100644 entry/src/main/resources/zh/element/string.json
55.  create mode 100644 entry/src/ohosTest/java/com/example/smartagriculture/ExampleOhosTest.java
56.  create mode 100644 entry/src/test/java/com/example/smartagriculture/ExampleTest.java
57.  create mode 100644 gradle.properties
58.  create mode 100644 gradle/wrapper/gradle-wrapper.jar
59.  create mode 100644 gradle/wrapper/gradle-wrapper.properties
60.  create mode 100644 gradlew
61.  create mode 100644 gradlew.bat
62.  create mode 100644 package.json
63.  create mode 100644 settings.gradle
64.
65.  taolee@DESKTOP-0HJVM6A MINGW64 ~/works/HarmonyOS/Books/SmartAgriculture (master)
66.  $ git status
67.  On branch master
68.  nothing to commit, working tree clean
```

第 2 行，通过 "git status" 命令查看当前项目的状态，发现任务新增的文件。

第 20～29 行，通过 "git add ." 命令将当前目录下的变动加入到暂存区，其中 "warning: LF will be replaced by CRLF in gradlew.bat." 表示文件行末尾换行符被自动修改，主要原因是 Linux/UNIX 操作系统和 Windows 操作系统对文件末尾的换行符不一样。

第 31 行，通过 "git commit -m "更改 App 图标和名称"" 命令提交变动到本地仓库。

第 66 行，通过 "git status" 命令查看当前项目目录已经干净，无改动。

2.8.3 将该版本代码打上标签

操作步骤如下

```
1.  taolee@DESKTOP-0HJVM6A MINGW64 ~/works/HarmonyOS/Books/SmartAgriculture (master)
2.  $ git tag -a task1 -m "更改 App 图标和名称"
3.  taolee@DESKTOP-0HJVM6A MINGW64 ~/works/HarmonyOS/Books/SmartAgriculture (master)
4.  $ git log --pretty=oneline
5.  4df17b7832e0746c1b154f22d34fea37069514c2 (HEAD -> master, tag: task1) 更改 App 图标和名称
6.  c8775743058113b200cb2e20a370e25a9056b365 Initial commit
```

第 2 行，通过 "git tag -a 标签名 -m 注释" 命令将当前分支最新提交打上标签。

第 5 行，通过 "git log --pretty=oneline" 命令打印日志，可以看到仓库最新版本处有标签 task1，即任务 1 的版本。

任务 3　创建 Splash 界面

任务概述

　　任务 1 已经构建了一个基本的 HarmonyOS 应用，而且完成了它在本地模拟器、远程模拟器、远程真机和本地真机中的运行。本任务要为应用创建一个 Splash 界面，如图 3-1 所示。Splash 界面主要用于显示应用的基本信息，是应用启动运行的第一个界面，停留 6 秒后自动或手动单击取消跳转到应用的登录界面（此处暂时在登录界面输出"Hello HarmonyOS"）。

图 3-1　Splash 界面

知识目标

- 了解 HarmonyOS 应用的工程目录。
- 掌握 Ability 的概念。
- 理解 Ability、Page Ability 和 Ability Slice。
- 掌握基本布局的概念。
- 了解 HarmonyOS 应用工程目录的关键文件。
- 了解 ResourceTable 类。

技能目标

- 能创建、编辑 Ability 和布局。
- 能处理组件的单击事件。
- 能使用 Ability 实现不同页面跳转。
- 能进行基本的布局设计。

3.1　HarmonyOS 应用的基础知识

　　HarmonyOS 应用的基础知识包括用户应用程序、用户应用程序包结构和关键术语。

3.1.1 用户应用程序

用户应用程序泛指运行在设备的操作系统之上，为用户提供特定服务的程序，简称"应用"。在 HarmonyOS 上运行的应用，有两种形态。
1）传统方式的需要安装的应用。
2）提供特定功能，免安装的应用（即原子化服务）。
在本书中，如无特殊说明，"应用"所指代的对象为上述需要安装的形态。

3.1.2 用户应用程序包结构

HarmonyOS 的用户应用程序包以 App Pack（Application Package）形式发布，它由一个或多个 HAP（HarmonyOS Ability Package）以及描述每个 HAP 属性的 pack.info 组成。HAP 是 Ability 的部署包，HarmonyOS 应用代码围绕 Ability 组件展开。一个 HAP 是由代码、资源、第三方库及应用配置文件组成的模块包，可分为 entry 和 feature 两种模块类型，如图 3-2 所示。

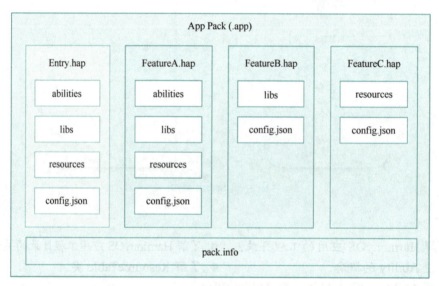

图 3-2　App Pack 结构

1．entry

应用的主模块。一个 App 中，对于同一设备类型，可以有一个或多个 entry 类型的 HAP，来支持该设备类型中不同规格（如 API 版本、屏幕规格等）的具体设备。如果同一设备类型存在多个 entry 模块，则必须配置 distroFilter 分发规则，使得应用市场在做应用的云端分发时，可实现对该设备类型下不同规格的设备进行精确分发。

2．feature

应用的动态特性模块。一个 App 可以包含一个或多个 feature 类型的 HAP，也可以不包含。只有包含 Ability 的 HAP 才能够独立运行。

3.1.3 关键术语

1. Ability

Ability 是应用所具备的能力的抽象，一个应用可以包含一个或多个 Ability。Ability 分为两种类型：FA（Feature Ability）和 PA（Particle Ability）。FA 和 PA 是应用的基本组成单元，能够实现特定的业务功能。FA 有 UI 界面，而 PA 无 UI 界面。

2. 库文件

库文件是应用依赖的第三方代码（例如 so、jar、bin、har 等二进制文件），存放在 libs 目录。

3. 资源文件

应用的资源文件（字符串、图片、音频等）存放于 resources 目录下，便于开发者使用和维护。

4. 配置文件

配置文件（config.json）是应用的 Ability 信息，用于声明应用的 Ability，以及应用所需权限等信息，详见应用配置文件。

5. pack.info

描述应用软件包中每个 HAP 的属性，由 IDE（集成开发环境）编译生成，应用市场根据该文件进行拆包和 HAP 的分类存储。HAP 的具体属性如下。

1）delivery-with-install：表示该 HAP 是否支持随应用安装。"true"表示支持随应用安装；"false"表示不支持随应用安装。
2）name：HAP 文件名。
3）module-type：模块类型，entry 或 feature。
4）device-type：表示支持该 HAP 运行的设备类型。

6. HAR

HAR（HarmonyOS Ability Resources）可以提供构建应用所需的所有内容，包括源代码、资源文件和 config.json 文件。HAR 不同于 HAP，HAR 不能独立安装运行在设备上，只能作为应用模块的依赖项被引用。

3.2 HarmonyOS 应用的配置文件

应用的每个 HAP 的根目录下都存在一个 config.json 配置文件，文件内容主要涵盖以下三个方面。

1）应用的全局配置信息，包含应用的包名、生产厂商、版本号等基本信息。
2）应用在具体设备上的配置信息，包含应用的备份恢复、网络安全等能力。
3）HAP 包的配置信息，包含每个 Ability 必须定义的基本属性（如包名、类名、类型以及 Ability 提供的能力），以及应用访问系统或其他应用受保护部分所需的权限等。

3.2.1 配置文件的组成

配置文件 config.json 采用 JSON 文件格式，其中包含了一系列配置项，每个配置项由属性和值两部分构成：属性和值。

1. 属性

属性出现顺序不分先后，且每个属性最多只允许出现一次。

2. 值

每个属性的值为 JSON 文件格式的基本数据类型（数值、字符串、布尔值、数组、对象或者 null 类型）。

3.2.2 配置文件的元素

DevEco Studio 提供了两种编辑 config.json 文件的方式。在 config.json 的编辑窗口中，可在右上角切换代码编辑视图，如图 3-3 所示。

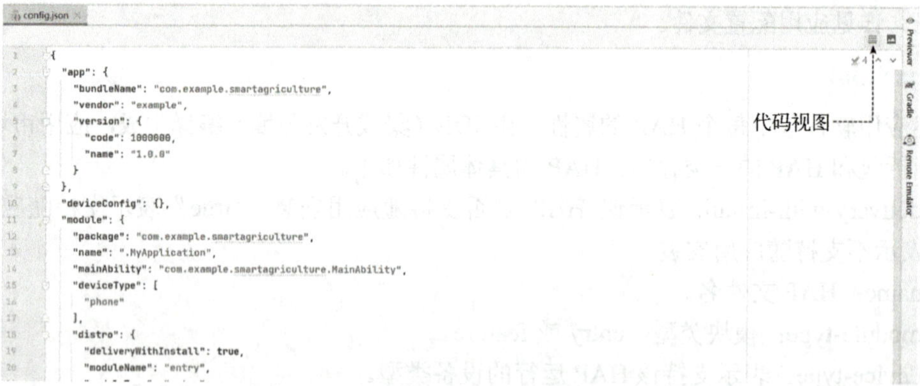

图 3-3　配置文件代码视图

也可以切换为可视化编辑视图，如图 3-4 所示。

图 3-4　配置文件可视化视图

3.2.3 配置文件内部结构

config.json 由 app、deviceConfig 和 module 三个部分组成,缺一不可,如表 3-1 所示。

表 3-1 配置文件的内部结构说明

属性名称	含 义	数据类型	是否可缺省
app	表示应用的全局配置信息。同一个应用的不同 HAP 包的 app 配置必须保持一致	对象	否
deviceConfig	表示应用在具体设备上的配置信息	对象	否
module	表示 HAP 包的配置信息。该标签下的配置只对当前 HAP 包生效	对象	否

3.2.4 app 对象的内部结构

app 对象包含应用的全局配置信息,如表 3-2 所示。

表 3-2 app 对象的内部结构说明

属性名称	子属性名称	含 义	数据类型	是否可缺省
bundleName	-	表示应用的包名,用于标识应用的唯一性。包名是由字母、数字、下画线(_)和点号(.)组成的字符串,必须以字母开头。支持的字符串长度为 7~127B 包名通常采用反域名形式表示(例如,com.huawei.himusic)。建议第一级为域名后缀 "com",第二级为厂商或个人名,第三级为应用名,也可以采用多级 说明: 如需使用 ohos.data.orm 包的接口,则应用的包名不能使用大写字母	字符串	否
vendor	-	表示对应用开发厂商的描述。字符串长度不超过 255B	字符串	可缺省,缺省值为空
version	-	表示应用的版本信息	对象	否
version	name	表示应用的版本号,用于向应用的终端用户呈现。取值可以自定义,长度不超过 127B。自定义规则如下 API 5 及更早版本:推荐使用三段式数字版本号(也兼容两段式版本号),如 A.B.C(也兼容 A.B),其中 A、B、C 取值为 0~999 范围内的整数。除此之外不支持其他格式 A 段:一般表示主版本号(Major) B 段:一般表示次版本号(Minor) C 段:一般表示修订版本号(Patch) API 6 版本起:推荐采用四段式数字版本号,如 A.B.C.D,其中 A、B、C 取值为 0~99 范围内的整数,D 取值为 0~999 范围内的整数 A 段:一般表示主版本号(Major) B 段:一般表示次版本号(Minor) C 段:一般表示特性版本号(Feature) D 段:一般表示修订版本号(Patch)	字符串	否
version	code	表示应用的版本号,仅用于 HarmonyOS 管理该应用,不对应用的终端用户呈现。取值规则如下 API 5 及更早版本:二进制 32 位以内的非负整数,需要从 version.name 的值转换得到 转换规则为:code 值=A×1,000,000 + B×1,000 + C 例如,version.name 字段取值为 2.2.1,则 code 的值为 2002001 API 6 版本起:code 的取值不与 version.name 字段的取值关联,开发者可自定义 code 的取值,取值范围为小于 2^{31} 的非负整数,但是应用的每次版本更新,均需更新 code 字段的值,新版本 code 取值必须大于旧版本 code 的值	数值	否
version	minCompatibleVersionCode	表示应用可兼容的最低版本号,用于在跨设备场景下,判断其他设备上该应用的版本是否兼容 格式与 version.code 字段的格式要求相同	数值	可缺省,缺省值为 code 标签值

（续）

属性名称	子属性名称	含义	数据类型	是否可缺省
smartWindowSize	-	该标签用于表示在悬浮窗场景下应用的模拟窗口的尺寸 配置格式为"正整数*正整数"，单位为 vp 正整数取值范围为[200,2000]	字符串	可缺省，缺省值为空
smartWindowDeviceType	-	表示应用可以在哪些设备上使用模拟窗口打开。取值如下 智能手机：phone 平板计算机：tablet 智慧屏：tv	字符串数组	可缺省，缺省值为空
targetBundleList	-	表示允许以免安装方式拉起①的其他 HarmonyOS 应用，列表取值为每个 HarmonyOS 应用的 bundleName，多个 bundleName 之间用英文","区分，最多配置 10 个 bundleName 如果被拉起的应用不支持免安装方式，则拉起失败	字符串	可缺省，缺省值为空

APP 示例如下。

```
1.  "app": {
2.    "bundleName": "com.example.smartagriculture",
3.    "vendor": "example",
4.    "version": {
5.      "code": 1000000,
6.      "name": "1.0.0"
7.    }
8.  }
```

3.2.5 deviceConfig 对象的内部结构

deviceConfig 包含在具体设备上的应用配置信息，可以包含 default、phone、tablet、tv、car、wearable、liteWearable 和 smartVision 等属性。default 标签内的配置适用于所有设备，其他设备类型如果有特殊的需求，则需要在该设备类型的标签下进行配置，deviceConfig 对内部结构说明如表 3-3 所见。

表 3-3　deviceConfig 对象的内部结构说明

属性名称	含义	数据类型	是否可缺省
default	表示所有设备通用的应用配置信息	对象	否
phone	表示手机类设备的应用信息配置	对象	可缺省，缺省为空
tablet	表示平板的应用配置信息	对象	可缺省，缺省为空
tv	表示智慧屏特有的应用配置信息	对象	可缺省，缺省为空
car	表示车机特有的应用配置信息	对象	可缺省，缺省为空
wearable	表示智能穿戴设备特有的应用配置信息	对象	可缺省，缺省为空
liteWearable	表示轻量级智能穿戴设备特有的应用配置信息	对象	可缺省，缺省为空
smartVision	表示智能摄像头特有的应用配置信息	对象	可缺省，缺省为空

default、phone、tablet、tv、car、wearable、liteWearable 和 smartVision 等对象的内部结构

---

① "拉起"即打开其他 App，是通过鸿蒙系统请求打开其他 App 页面，区别于某个 App 打开自己的页面。

说明如表 3-4 所示。

表 3-4 不同设备的内部结构说明

属性名称	含 义	数据类型	是否可缺省
jointUserId	表示应用的共享 UserId 通常情况下，不同的应用运行在不同的进程中，应用的资源无法共享。如果开发者的多个应用之间需要共享资源，则可以通过相同的 jointUserId 值实现，前提是这些应用的签名相同 该标签仅对系统应用生效，且仅适用于手机、平板计算机、智慧屏、车机（车载计算机）、智能穿戴 该字段在 API Version 3 及更高版本不再支持配置设备	字符串	可缺省，缺省为空
process	表示应用或者 Ability 的进程名 如果在 deviceConfig 标签下配置了 process 标签，则该应用的所有 Ability 都运行在这个进程中。如果在 abilities 标签下也为某个 Ability 配置了 process 标签，则该 Ability 就运行在这个进程中 该标签仅适用于手机、平板计算机、智慧屏、车机、智能穿戴设备	字符串	可缺省，缺省为应用的软件包名
supportBackup	表示应用是否支持备份和恢复。如果配置为 "false"，则不支持为该应用执行备份或恢复操作 该标签仅适用于手机、平板计算机、智慧屏、车机、智能穿戴设备	布尔类型	可缺省，缺省为 "false"
compressNativeLibs	表示 libs 库是否以压缩存储的方式打到 HAP 包。如果配置为 "false"，则 libs 库以不压缩的方式存储，HAP 包在安装时无须解压 libs，运行时会直接从 HAP 内加载 libs 库 该标签仅适用于手机、平板计算机、智慧屏、车机、智能穿戴设备	布尔类型	可缺省，缺省为 "true"
network	表示网络安全性配置。该标签允许应用通过配置文件的安全声明来自定义其网络安全，无需修改应用代码	对象	可缺省，缺省为空

network 对象内部结构如表 3-5 所示。

表 3-5 network 对象的内部结构说明

属性名称	含 义	数据类型	是否可缺省
cleartextTraffic	表示是否允许应用使用明文网络流量（例如，明文 HTTP） true：允许应用使用明文流量的请求 false：拒绝应用使用明文流量的请求	布尔类型	可缺省，缺省为 "false"
securityConfig	表示应用的网络安全配置信息	对象	可缺省，缺省为空

securityConfig 对象内部结构如表 3-6 所示。

表 3-6 securityConfig 对象的内部结构说明

属性名称	子属性名称	含 义	数据类型	是否可缺省
domainSettings	-	表示自定义的网域范围的安全配置，支持多层嵌套，即一个 domainSettings 对象中允许嵌套更小网域范围的 domainSettings 对象	对象	可缺省，缺省为空
	cleartextPermitted	表示自定义的网域范围内是否允许明文流量传输。当 cleartextTraffic 和 securityConfig 同时存在时，自定义网域是否允许明文流量传输以 cleartextPermitted 的取值为准 true：允许明文流量传输 false：拒绝明文流量传输	布尔类型	否
	domains	表示域名配置信息，包含两个参数：subdomains 和 name subdomains（布尔类型）：表示是否包含子域名。如果为 "true"，此网域规则将与相应网域及所有子域（包括子网域的子网域）匹配。否则，该规则仅适用于精确匹配项 name（字符串）：表示域名名称	对象数组	否

deviceConfig 示例如下。

```
1.  "deviceConfig": {
2.    "default": {
3.      "process": "com.huawei.hiworld.example",
```

```
4.         "supportBackup": false,
5.         "network": {
6.             "cleartextTraffic": true,
7.             "securityConfig": {
8.                 "domainSettings": {
9.                     "cleartextPermitted": true,
10.                    "domains": [
11.                        {
12.                            "subdomains": true,
13.                            "name": "example.ohos.com"
14.                        }
15.                    ]
16.                }
17.            }
18.        }
19.    }
20. }
```

3.2.6　module 对象的内部结构

module 对象包含 HAP 包的配置信息，内部结构说明如表 3-7 所示。

表 3-7　module 对象的内部结构说明

属性名称	含　　义	数据类型	是否可缺省
mainAbility	表示 HAP 包的入口 Ability 名称。该标签的值应配置为"module > abilities"中存在的 Page 类型 Ability 的名称。该标签仅适用于手机、平板、智慧屏、车机、智能穿戴设备	字符串	如果存在 page 类型的 ability，则该字段不可缺省
package	表示 HAP 的包结构名称，在应用内应保证唯一性。采用反向域名格式（建议与 HAP 的工程目录保持一致）。字符串长度不超过 127B。该标签仅适用于手机、平板计算机、智慧屏、车机、智能穿戴设备	字符串	否
name	表示 HAP 的类名。采用反向域名方式表示，前缀需要与同级的 package 标签指定的包名一致，也可采用"."开头的命名方式。字符串长度不超过 255B。该标签仅适用于手机、平板计算机、智慧屏、车机、智能穿戴设备	字符串	否
description	表示 HAP 的描述信息。字符串长度不超过 255B。如果字符串超出长度或者需要支持多语言，可以采用资源索引的方式添加描述内容。该标签仅适用于手机、平板计算机、智慧屏、车机、智能穿戴设备	字符串	可缺省，缺省值为空
supportedModes	表示应用支持的运行模式。当前只定义了驾驶模式（drive）。该标签仅适用于车机	字符串数组	可缺省，缺省值为空
deviceType	表示允许 Ability 运行的设备类型。系统预定义的设备类型包括：phone（手机）、tablet（平板）、tv（智慧屏）、car（车机）、wearable（智能穿戴设备）、liteWearable（轻量级智能穿戴设备）等	字符串数组	否
distro	表示 HAP 发布的具体描述。该标签仅适用于手机、平板计算机、智慧屏、车机、智能穿戴设备	对象	否
metaData	表示 HAP 的元信息	对象	可缺省，缺省值为空
abilities	表示当前模块内的所有 Ability。采用对象数组格式，其中每个元素表示一个 Ability 对象	对象数组	可缺省，缺省值为空
js	表示基于 JS UI 框架开发的 JS 模块集合，其中的每个元素代表一个 JS 模块的信息	对象数组	可缺省，缺省值为空
shortcuts	表示应用的快捷方式信息。采用对象数组格式，其中每个元素表示一个快捷方式对象	对象数组	可缺省，缺省值为空
defPermissions	表示应用定义的权限。应用调用者必须申请这些权限，才能正常调用该应用	对象数组	可缺省，缺省值为空

（续）

属性名称	含 义	数据类型	是否可缺省
reqPermissions	表示应用运行时向系统申请的权限	对象数组	可缺省，缺省值为空
colorMode	表示应用自身的颜色模式 dark：表示按照深色模式选取资源 light：表示按照浅色模式选取资源 auto：表示跟随系统的颜色模式值选取资源。该标签仅适用于手机、平板、智慧屏、车机、智能穿戴设备	字符串	可缺省，缺省值为"auto"
resizeable	表示应用是否支持多窗口特性。该标签仅适用于手机、平板计算机、智慧屏、车机、智能穿戴设备	布尔类型	可缺省，缺省值为"true"
distroFilter	表示应用的分发规则。该标签用于定义 HAP 包对应的细分设备规格的分发策略，以便在应用市场进行云端分发应用包时做精准匹配。该标签可配置的分发策略维度包括 API Version、屏幕形状、屏幕分辨率。在进行分发时，通过 deviceType 与这三个属性的匹配关系，唯一确定一个用于分发到对应设备的 HAP	对象数组	可缺省。缺省值为空。但当应用中包含多个 entry 模块时，必须配置该标签

module 示例如下。

```
1.   "module": {
2.       "package": "com.example.smartagriculture",
3.       "name": ".MyApplication",
4.       "mainAbility": "com.example.smartlawn.MainAbility",
5.       "deviceType": [
6.           "phone",
7.           "tablet"
8.       ],
9.       "distro": {
10.          "deliveryWithInstall": true,
11.          "moduleName": "entry",
12.          "moduleType": "entry",
13.          "installationFree": false
14.      },
15.      "abilities": [
16.          ...
17.      ],
18.      "shortcuts": [
19.          ...
20.      ],
21.      "js": [
22.          ...
23.      ],
24.      "reqPermissions": [
25.          ...
26.      ],
27.      "defPermissions": [
28.          ...
29.      ],
30.      "colorMode": "light"
31.  }
```

distro 对象的内部结构说明如表 3-8 所示。

表 3-8　distro 对象的内部结构说明

属性名称	含义	数据类型	是否可缺省
deliveryWithInstall	表示当前 HAP 是否支持随应用安装 true：支持随应用安装 false：不支持随应用安装 该属性建议设置为 true 若设置 false 可能导致最终应用上架应用市场时发生异常	布尔类型	否
moduleName	表示当前 HAP 的名称	字符串	否
moduleType	表示当前 HAP 的类型，包括两种类型：entry 和 feature	字符串	否
installationFree	表示当前 HAP 是否支持免安装特性 true：表示支持免安装特性，且符合免安装约束 false：表示不支持免安装特性 另外还需注意： 当 entry.hap 的字段配置为 true 时，与该 entry.hap 相关的所有 feature.hap 的字段也需要配置为 true 当 entry.hap 的字段配置为 false 时，与该 entry.hap 相关的各 feature.hap 的字段可按业务需求配置 true 或 false	布尔类型	否

metaData 对象的内部结构说明如表 3-9 所示。

表 3-9　metaData 对象的内部结构说明

属性名称	子属性名称	含义	数据类型	是否可缺省
parameters	-	表示调用 Ability 时所有调用参数的元信息。每个调用参数的元信息由以下三个标签组成：description、name、type	对象	可缺省，缺省值为空
	description	表示对调用参数的描述，可以是表示描述内容的字符串，也可以是对描述内容的资源索引，以支持多语言	字符串	可缺省，缺省值为空
	name	表示调用参数的名称	字符串	可缺省，缺省值为空
	type	表示调用参数的类型，如 Integer	字符串	否
results	-	表示 Ability 返回值的元信息。每个返回值的元信息由以下三个标签组成：description、name、type	对象	可缺省，缺省值为空
	description	表示对返回值的描述，可以是表示描述内容的字符串，也可以是对描述内容的资源索引以支持多语言	字符串	可缺省，缺省值为空
	name	表示返回值的名字	字符串	可缺省，缺省值为空
	type	表示返回值的类型，如 Integer	字符串	否
customizeData	-	表示父级组件的自定义元信息，parameters 和 results 在 module 中不可配	对象	可缺省，缺省值为空
	name	表示数据项的键名称，字符串类型（最大长度 255 字节）	字符串	可缺省，缺省值为空
	value	表示数据项的值，字符串类型（最大长度 255 字节）	字符串	可缺省，缺省值为空
	extra	表示用户自定义数据格式，标签值为标识该数据的资源索引值	字符串	可缺省，缺省值为空

abilities 对象的内部结构说明如表 3-10 所示。

表 3-10　abilities 对象的内部结构说明

属性名称	含义	数据类型	是否可缺省
name	表示 Ability 名称。取值可采用反向域名方式表示，由包名和类名组成，如"com.example.myapplication.MainAbility"；也可采用"."开头的类名方式表示，如".MainAbility"。Ability 的名称，需在一个应用的范围内保证唯一。该标签仅适用于手机、平板计算机、智慧屏、车机、智能穿戴设备 在使用 DevEco Studio 新建项目时，默认生成首个 Ability 的配置，包括生成"MainAbility.java"文件，及"config.json"中"MainAbility"的配置。如使用其他 IDE 工具，可自定义名称	字符串	否

（续）

属性名称	含 义	数据类型	是否可缺省
description	表示对 Ability 的描述。取值可以是描述性内容，也可以是对描述性内容的资源索引，以支持多语言	字符串	可缺省，缺省值为空
icon	表示 Ability 图标资源文件的索引。取值示例：$media:ability_icon。如果在该 Ability 的 skills 属性中，actions 的取值包含"action.system.home"，entities 取值中包含"entity.system.home"，则该 Ability 的 icon 将同时作为应用的 icon。如果存在多个符合条件的 Ability，则取位置靠前的 Ability 的 icon 作为应用的 icon 应用的"icon"和"label"是用户可感知配置项，需要区别于当前所有已有的应用"icon"或"label"（至少有一个不同）	字符串	可缺省，缺省值为空
label	表示 Ability 对用户显示的名称。取值可以是 Ability 名称，也可以是对该名称的资源索引，以支持多语言 如果在该 Ability 的 skills 属性中，actions 的取值包含"action.system.home"，entities 取值中包含"entity.system.home"，则该 Ability 的 label 将同时作为应用的 label。如果存在多个符合条件的 Ability，则取位置靠前的 Ability 的 label 作为应用的 label 应用的"icon"和"label"是用户可感知配置项，需要区别于当前所有已有的应用"icon"或"label"（至少有一个不同）	字符串	可缺省，缺省值为空
uri	表示 Ability 的统一资源标识符 格式为[scheme:][//authority][path][?query][#fragment]	字符串	可缺省，对于 data 类型的 Ability 不可缺省
launchType	表示 Ability 的启动模式，支持"standard"、"singleMission"和"singleton"三种模式 standard：表示该 Ability 可以有多实例。"standard"模式适用于大多数应用场景 singleMission：表示此 Ability 在每个任务栈中只能有一个实例 singleton：表示该 Ability 在所有任务栈中仅可以有一个实例。例如，具有全局唯一性的呼叫来电界面即采用"singleton"模式 该标签仅适用于手机、平板计算机、智慧屏、车机、智能穿戴设备	字符串	可缺省，缺省值为"standard"
visible	表示 Ability 是否可以被其他应用调用 true：可以被其他应用调用 false：不能被其他应用调用	布尔类型	可缺省，缺省值为"false"
permissions	表示其他应用的 Ability 调用此 Ability 时需要申请的权限。通常采用反向域名格式，取值可以是系统预定义的权限，也可以是开发者自定义的权限。如果是自定义权限，取值必须与 defPermissions 标签中定义的某个权限的 name 标签值一致	字符串数组	可缺省，缺省值为空
skills	表示 Ability 能够接收的 Intent 的特征	对象数组	可缺省，缺省值为空
deviceCapability	表示 Ability 运行时要求设备具有的能力，采用字符串数组的格式表示	字符串数组	可缺省，缺省值为空
metaData	表示 Ability 的元信息。 调用 Ability 时调用参数的元信息，例如：参数个数和类型 Ability 执行完毕返回值的元信息，例如：返回值个数和类型 该标签仅适用于智慧屏、智能穿戴设备、车机	对象	可缺省，缺省值为空
type	表示 Ability 的类型。取值范围如下： page：表示基于 Page 模板开发的 FA，用于提供与用户交互的能力 service：表示基于 Service 模板开发的 PA，用于提供后台运行任务的能力 data：表示基于 Data 模板开发的 PA，用于对外部提供统一的数据访问抽象 CA：表示支持其他应用以窗口方式调起该 Ability	字符串	否
orientation	表示该 Ability 的显示模式。该标签仅适用于 page 类型的 Ability。取值范围如下 unspecified：由系统自动判断显示方向 landscape：横屏模式 portrait：竖屏模式 followRecent：跟随栈中最近的应用	字符串	可缺省，缺省值为"unspecified"
backgroundModes	表示后台服务的类型，可以为一个服务配置多个后台服务类型。该标签仅适用于 service 类型的 Ability。取值范围如下 dataTransfer：通过网络或对端设备进行数据下载、备份、分享、传输等业务 audioPlayback：音频输出业务 audioRecording：音频输入业务 pictureInPicture：画中画、小窗口播放视频业务	字符串数组	可缺省，缺省值为空

（续）

属性名称	含 义	数据类型	是否可缺省
backgroundModes	voip：音视频电话、VOIP 业务 location：定位、导航业务 bluetoothInteraction：蓝牙扫描、连接、传输业务 wifiInteraction：WLAN 扫描、连接、传输业务 screenFetch：录屏、截屏业务 multiDeviceConnection：多设备互联业务	字符串数组	可缺省，缺省值为空
readPermission	表示读取 Ability 的数据所需的权限。该标签仅适用于 data 类型的 Ability 取值为长度不超过 255 字节的字符串 该标签仅适用于手机、平板计算机、智慧屏、车机、智能穿戴设备	字符串	可缺省，缺省为空
writePermission	表示向 Ability 写数据所需的权限。该标签仅适用于 data 类型的 Ability。取值为长度不超过 255 字节的字符串 该标签仅适用于手机、平板计算机、智慧屏、车机、智能穿戴设备	字符串	可缺省，缺省为空
configChanges	表示 Ability 关注的系统配置集合。当已关注的配置发生变更后，Ability 会收到 onConfigurationUpdated 回调。取值范围如下 mcc：表示 IMSI 移动设备国家/地区代码（MCC）发生变更。典型场景：检测到 SIM 并更新 MCC mnc：IMSI 移动设备网络代码（MNC）发生变更。典型场景：检测到 SIM 并更新 MNC locale：表示语言区域发生变更。典型场景：用户已为设备文本的文本显示选择新的语言类型 layout：表示屏幕布局发生变更。典型场景：当前有不同的显示形态都处于活跃状态 fontSize：表示字号发生变更。典型场景：用户已设置新的全局字号 orientation：表示屏幕方向发生变更。典型场景：用户旋转设备 density：表示显示密度发生变更。典型场景：用户可能指定不同的显示比例，或当前有不同的显示形态同时处于活跃状态 size：显示窗口大小发生变更 smallestSize：显示窗口较短边的边长发生变更 colorMode：颜色模式发生变更	字符串数组	可缺省，缺省为空
mission	表示 Ability 指定的任务栈。该标签仅适用于 page 类型的 Ability。默认情况下应用中所有 Ability 同属一个任务栈 该标签仅适用于手机、平板计算机、智慧屏、车机、智能穿戴设备	字符串	可缺省，缺省为应用的包名
targetAbility	表示当前 Ability 重用的目标 Ability。该标签仅适用于 page 类型的 Ability。如果配置了 targetAbility 属性，则当前 Ability（即别名 Ability）的属性中仅 name、icon、label、visible、permissions、skills 生效，其他属性均沿用 targetAbility 中的属性值。目标 Ability 必须与别名 Ability 在同一应用中，且在配置文件中目标 Ability 必须在别名之前进行声明 该标签仅适用于手机、平板计算机、智慧屏、车机、智能穿戴设备	字符串	可缺省，缺省值为空。表示当前 Ability 不是一个别名 Ability
multiUserShared	表示 Ability 是否支持多用户状态进行共享，该标签仅适用于 data 类型的 Ability 配置为"true"时，表示在多用户下只有一份存储数据。需要注意的是，该属性会使 visible 属性失效 该标签仅适用于手机、平板计算机、智慧屏、车机、智能穿戴设备	布尔类型	可缺省，缺省值为"false"
supportPipMode	表示 Ability 是否支持用户进入 PIP 模式（用于在页面最上层悬浮小窗口，俗称"画中画"，常见于视频播放等场景）。该标签仅适用于 page 类型的 Ability 该标签仅适用于手机、平板计算机、智慧屏、车机、智能穿戴设备	布尔类型	可缺省，缺省值为"false"
formsEnabled	表示 Ability 是否支持卡片（forms）功能。该标签仅适用于 page 类型的 Ability true：支持卡片能力 false：不支持卡片能力	布尔类型	可缺省，缺省值为"false"
forms	表示服务卡片的属性。该标签仅当 formsEnabled 为"true"时，才能生效	对象数组	可缺省，缺省值为空
resizeable	表示 Ability 是否支持多窗口特性 该标签仅适用于手机、平板计算机、智慧屏、车机、智能穿戴设备	布尔类型	可缺省，缺省值为"true"

abilities 示例如下。

1. "abilities": [

```
2.     {
3.       "skills": [
4.         {
5.           "entities": [
6.             "entity.system.home"
7.           ],
8.           "actions": [
9.             "action.system.home"
10.          ]
11.        }
12.      ],
13.      "orientation": "unspecified",
14.      "name": "com.example.smartlawn.MainAbility",
15.      "icon": "$media:logo",
16.      "description": "$string:mainability_description",
17.      "label": "$string:app_name",
18.      "type": "page",
19.      "launchType": "standard"
20.    }
21.  ]
```

skills 对象的内部结构说明如表 3-11 所示。

表 3-11　skills 对象的内部结构说明

属性名称	子属性名称	含　义	数据类型	是否可缺省
actions	-	表示能够接收的 Intent 的 action 值，可以包含一个或多个 action 取值通常为系统预定义的 action 值，详见 HarmonyOS 官网《API 参考》中的 ohos.aafwk.content.Intent 类	字符串数组	可缺省，缺省值为空
entities	-	表示能够接收的 Intent 的 Ability 的类别（如视频、桌面应用等），可以包含一个或多个 entity 取值通常为系统预定义的类别，详见 HarmonyOS 官网《API 参考》中的 ohos.aafwk.content.Intent 类，也可以自定义	字符串数组	可缺省，缺省值为空
uris	-	表示能够接收的 Intent 的 uri，可以包含一个或多个 uri	对象数组	可缺省，缺省值为空
	scheme	表示 uri 的 scheme 值	字符串	不可缺省
	host	表示 uri 的 host 值	字符串	可缺省，缺省值为空
	port	表示 uri 的 port 值	字符串	可缺省，缺省值为空
	path	表示 uri 的 path 值	字符串	可缺省，缺省值为空
	type	表示 uri 的 type 值	字符串	可缺省，缺省值为空

skills 示例如下。

```
1.  "skills": [
2.    {
3.      "entities": [
4.        "entity.system.home"
5.      ],
6.      "actions": [
```

```
7.                "action.system.home"
8.            ]
9.        }
10.    ]
```

3.3　HarmonyOS 应用的资源文件

HarmonyOS 应用的资源文件包括字符串、图片、音频、布局文件等，所有资源文件都放置于资源目录之下，并进行分类管理。

3.3.1　resources 目录

应用的资源文件（字符串、图片、音频等）统一存放于 resources 目录下，便于开发者使用和维护。resources 目录包括两大类目录，一类为 base 目录与限定词目录，另一类为 rawfile 目录。资源目录示例如下。

```
1.  resources
2.  |---base                              // 默认存在的目录
3.  |   |---element
4.  |   |   |---string.json
5.  |   |---media
6.  |   |   |---icon.png
7.  |---en_GB-vertical-car-mdpi           // 限定词目录示例，需要开发者自行创建
8.  |   |---element
9.  |   |   |---string.json
10. |   |---media
11. |   |   |---icon.png
12. |---rawfile                           // 默认存在的目录
```

resources 目录分类如表 3-12 所示。

表 3-12　resources 目录分类

分类	base 目录与限定词目录	rawfile 目录
组织形式	按照两级目录形式来组织，目录命名必须符合规范，以便根据设备状态去匹配相应目录下的资源文件 一级子目录为 base 目录和限定词目录 base 目录是默认存在的目录。当应用的 resources 资源目录中没有与设备状态匹配的限定词目录时，会自动引用该目录中的资源文件 限定词目录需要开发者自行创建。目录名称由一个或多个表征应用场景或设备特征的限定词组合而成，具体要求参见限定词目录 二级子目录为资源目录，用于存放字符串、颜色、布尔值等基础元素，以及媒体、动画、布局等资源文件，具体要求参见资源组目录	支持创建多层子录，目录名称可以自定义，文件夹内可以自由放置各类资源文件 rawfile 目录的文件不会根据设备状态去匹配不同的资源
编译方式	目录中的资源文件会被编译成二进制文件，并赋予资源文件 ID	目录中的资源文件会被直接打包进应用，不经过编译，也不会被赋予资源文件 ID
引用方式	通过指定资源类型（type）和资源名称（name）来引用	通过指定文件路径和文件名来引用

3.3.2 限定词目录

限定词目录可以由一个或多个表征应用场景或设备特征的限定词组合而成，包括移动国家码（MCC）和移动网络码（MNC）、语言、文字、国家或地区、横竖屏、设备类型、颜色模式和屏幕密度等，限定词之间通过下画线（_）或者短横线（-）连接。开发者在创建限定词目录时，需要掌握限定词目录的命名要求，以及限定词目录与设备状态的匹配规则。

1. 限定词目录的命名要求

1）限定词的组合顺序：移动国家码_移动网络码-语言_文字_国家或地区-横竖屏-设备类型-颜色模式-屏幕密度。开发者可以根据应用的使用场景和设备特征，选择其中的一类或几类限定词组成目录名称。

2）限定词的连接方式：语言、文字、国家或地区之间、移动国家码和移动网络码之间采用下画线（_）连接，其他限定词之间均采用短横线（-）连接。例如：zh_Hant_CN、zh_CN-car-ldpi。

3）限定词的取值范围：每类限定词的取值必须符合表 3-13 中的条件，否则，将无法匹配目录中的资源文件。

表 3-13　限定词取值要求

限定词类型	含义与取值说明
移动国家码和移动网络码	移动国家码（MCC）和移动网络码（MNC）的值取自设备注册的网络。MCC 后面可以跟随 MNC，使用下画线（_）连接，也可以单独使用。例如：mcc460 表示中国，mcc460_mnc00 表示中国_中国移动。详细取值范围，请查阅 ITU-T E.212（国际电联相关建议书）
语言	表示设备使用的语言类型，由 2~3 个小写字母组成。例如：zh 表示中文，en 表示英语，mai 表示迈蒂利语。详细取值范围，请查阅 ISO 639（ISO 制定的语言编码标准，GB/T 4880.2-2000）
文字	表示设备使用的文字类型，由 1 个大写字母（首字母）和 3 个小写字母组成。例如：Hans 表示简体中文，Hant 表示繁体中文。详细取值范围，请查阅 ISO 15924（ISO 制定的文字编码标准）
国家或地区	表示用户所在的国家或地区，由 2~3 个大写字母或者 3 个数字组成。例如：CN 表示中国，GB 表示英国。详细取值范围请查阅 ISO 3166-1（ISO 制定的国家和地区编码标准）
横竖屏	表示设备的屏幕方向，取值如下 vertical：竖屏 horizontal：横屏
设备类型	表示设备的类型，取值如下 phone：手机 tablet：平板计算机 car：车机 tv：智慧屏 wearable：智能穿戴设备
颜色模式	表示设备的颜色模式，取值如下 dark：深色模式 light：浅色模式
屏幕密度	表示设备的屏幕密度（单位为 dpi），取值如下 sdpi：表示小规模的屏幕密度，适用于 dpi 取值为(0, 120]的设备 mdpi：表示中规模的屏幕密度，适用于 dpi 取值为(120, 160]的设备 ldpi：表示大规模的屏幕密度，适用于 dpi 取值为(160, 240]的设备 xldpi：表示特大规模的屏幕密度，适用于 dpi 取值为(240, 320]的设备 xxldpi：表示超大规模的屏幕密度，适用于 dpi 取值为(320, 480]的设备 xxxldpi：表示超特大规模的屏幕密度（Extra Extra Extra Large-scale Dots Per Inch），适用于 dpi 取值为(480, 640]的设备

2. 限定词目录与设备状态的匹配规则

1）在为设备匹配对应的资源文件时，限定词目录匹配的优先级从高到低依次为：移动国家码和移动网络码>区域（可选组合：语言、语言_文字、语言_国家或地区、语言_文字_国家或地区）>横竖屏>设备类型>颜色模式>屏幕密度。

2）如果限定词目录中包含移动国家码和移动网络码、语言、文字、横竖屏、设备类型、颜色模式等限定词，则对应限定词的取值必须与当前的设备状态完全一致，该目录才能够参与设备的资源匹配。例如，限定词目录"zh_CN-car-ldpi"不能参与"en_US"设备的资源匹配。

3.3.3 资源组目录

base 目录与限定词目录下面可以创建资源组目录（包括 element、media、animation、layout、graphic、profile），用于存放特定类型的资源文件，如表 3-14 所示。

表 3-14 资源组目录说明

资源组目录	目录说明	资源文件
element	表示元素资源，以下每一类数据都采用相应的 JSON 文件来表征 boolean：布尔型 color：颜色 float：浮点型 intarray：整型数组 integer：整型 pattern：样式 plural：复数形式 strarray：字符串数组 string：字符串	element 目录中的文件名称建议与下面的文件名保持一致。每个文件中只能包含同一类型的数据 boolean.json color.json float.json intarray.json integer.json pattern.json plural.json strarray.json string.json
media	表示媒体资源，包括图片、音频、视频等非文本格式的文件	文件名可自定义，例如：icon.png
animation	表示动画资源，采用 XML 文件格式	文件名可自定义，例如：zoom_in.xml
layout	表示布局资源，采用 XML 文件格式	文件名可自定义，例如：home_layout.xml
graphic	表示可绘制资源，采用 XML 文件格式	文件名可自定义，例如：notifications_dark.xml
profile	表示其他类型文件，以原始文件形式保存	文件名可自定义

3.4 创建 Splash 界面

Splash 界面主要用于显示应用的基本信息，包括应用的 Logo、名称、版本等信息，该界面是应用启动运行的第一个界面，一般停留秒数后自动或手动跳转到应用的登录界面。

3.4.1 了解项目工程

打开任务 1 的项目工程，项目工程视图如图 3-5 所示，HarmonyOS 应用实际上就是使用文件夹结构组织的一系列文件的集合。DevEco Studio 在创建工程的时候会自动创建这些文件。

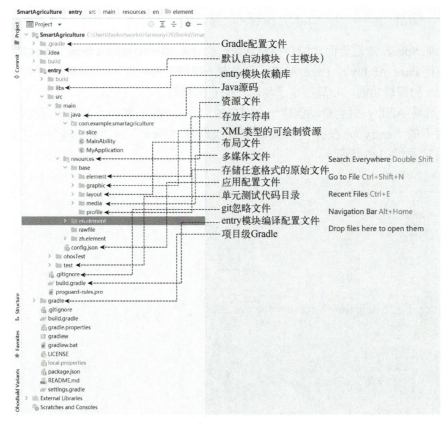

图 3-5　项目工程视图

3.4.2　了解 Ability 基础

Ability 是应用所具备能力的抽象，也是应用程序的重要组成部分。一个应用可以具备多种能力（即可以包含多个 Ability），HarmonyOS 支持应用以 Ability 为单位进行部署。Ability 可以分为 FA（Feature Ability）和 PA（Particle Ability）两种类型，每种类型为开发者提供了不同的模板，以便实现不同的业务功能。

1. FA 支持 Page Ability

Page 模板是 FA 唯一支持的模板，用于提供与用户交互的能力。一个 Page 实例可以包含一组相关页面，每个页面用一个 AbilitySlice 实例表示。

2. PA 支持 Service Ability 和 Data Ability

1）Service 模板：用于提供后台运行任务的能力。

2）Data 模板：用于对外部提供统一的数据访问抽象。

在配置文件（config.json）中注册 Ability 时，可以通过配置 Ability 元素中的"type"属性来指定 Ability 模板类型，"type"的取值可以为"page""service"或"data"，分别代表 Page 模板、Service 模板、Data 模板。

3.4.3 创建 Splash Ability 和布局

为了实现 Splash 欢迎界面功能，需要创建一个 FA 和一个布局（若本书不做特别说明，则 "Ability" 指代 Page Ability）。Page Ability 可以有一组与之相关的 AbilitySlice，SplashAbilitySlice 指定了 Splash 的逻辑功能，布局指定了与用户交互的界面。

一般，新增 Ability 时会自动创建一个对应的 AbilitySlice 和一个与之相应的布局文件。右击项目视图下的"entry"，在弹出的菜单中依次选择"New"→"Ability"→"Empyt Page Ability(Java)"，如图 3-6 所示。

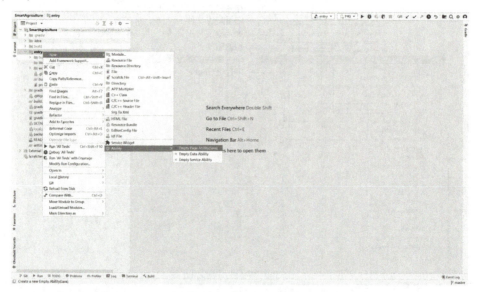

图 3-6　新增 Ability

如图 3-7 所示，在配置 Ability 的界面中，按照图中的序号，依次设置，完成 Ability 的创建。其中，Launcher Ability 表示自动设置该 Ability 为启动运行的第一个 Ability，也可以通过设置配置文件 config.json 完成。

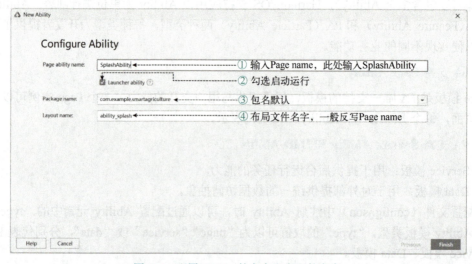

图 3-7　配置 Ability 的名字和布局的名字

如图 3-8 所示，DevEco Studio 自动产生了三个文件，其中 SplashAbility 表示 Page Ability，SplashAbilitySlice 表示 Page Ability 下面的一个 AbilitySlice，ability_splash.xml 表示 SplashAbilitySlice 对应的界面显示内容。

图 3-8 Splash Ability 和布局

3.4.4 编辑配置文件

编辑 config.json 文件中"module"→"abilities"→"SplashAbility"的 skills 元素。完整的 config.json 文件如下。

```
1.  {
2.    "app": {
3.      "bundleName": "com.example.smartagriculture",
4.      "vendor": "example",
5.      "version": {
6.        "code": 1000000,
7.        "name": "1.0.0"
8.      }
9.    },
10.   "deviceConfig": {},
11.   "module": {
12.     "package": "com.example.smartagriculture",
13.     "name": ".MyApplication",
14.     "mainAbility": "com.example.smartagriculture.MainAbility",
15.     "deviceType": [
16.       "phone"
17.     ],
18.     "distro": {
19.       "deliveryWithInstall": true,
20.       "moduleName": "entry",
```

```
21.        "moduleType": "entry",
22.        "installationFree": false
23.      },
24.      "abilities": [
25.        {
26.          "skills": [
27.          ],
28.          "orientation": "unspecified",
29.          "visible": true,
30.          "name": "com.example.smartagriculture.MainAbility",
31.          "icon": "$media:logo",
32.          "description": "$string:mainability_description",
33.          "label": "$string:app_name",
34.          "type": "page",
35.          "launchType": "standard"
36.        },
37.        {
38.          "skills": [
39.            {
40.              "entities": [
41.                "entity.system.home"
42.              ],
43.              "actions": [
44.                "action.system.home"
45.              ]
46.            }
47.          ],
48.          "orientation": "unspecified",
49.          "visible": true,
50.          "name": "com.example.smartagriculture.SplashAbility",
51.          "icon": "$media:logo",
52.          "description": "$string:splashability_description",
53.          "label": "$string:app_name",
54.          "type": "page",
55.          "launchType": "standard"
56.        }
57.      ]
58.    }
59. }
```

第 39~46 行，将 Splash 用于 HarmonyOS 运行的第一个页面。

第 51、53 行，根据表 3-10 中对 icon 和 label 的描述，如果此 Ability 中 skills 元素 actions 的取值包含"action.system.home"，entities 取值中包含"entity.system.home"，则该 Ability 的 icon 将同时作为应用的 icon，label 作为应用的 label。修改 icon 和 label 即达到修改应用的图标和名称的效果。

编译运行应用，可以得到如图 3-9 和图 3-10 所示结果。

任务 3　创建 Splash 界面

图 3-9　运行 App

图 3-10　App 图标和名称

3.4.5　编辑 Splash 布局

目前在 DevEco Studio 中增加 GUI 的方法只有通过在布局文件中以代码视图方式增加，如图 3-11 所示，双击打开 ability_splash.xml 文件，可以看到布局的代码视图区域，单击右侧的 Previewer，可以查看界面设计的预览效果。

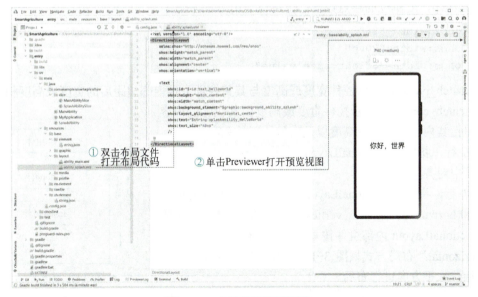

图 3-11　布局设计代码视图

布局文件代码如下。

```xml
1.  <?xml version="1.0" encoding="utf-8"?>
2.  <DirectionalLayout
3.      xmlns:ohos="http://schemas.huawei.com/res/ohos"
4.      ohos:height="match_parent"
5.      ohos:width="match_parent"
6.      ohos:alignment="center"
7.      ohos:orientation="vertical">
8.
9.      <Text
10.         ohos:id="$+id:text_helloworld"
11.         ohos:height="match_content"
12.         ohos:width="match_content"
13.         ohos:background_element="$graphic:background_ability_splash"
14.         ohos:layout_alignment="horizontal_center"
15.         ohos:text="$string:splashability_HelloWorld"
16.         ohos:text_size="40vp"
17.         />
18.
19. </DirectionalLayout>
```

第 1 行，xml 格式的文件的版本和编码的定义。

第 2 行，DirectionalLayout 是 Java UI 中的一种重要组件布局方式，用于将一组组件（Component）按照水平或者垂直方向排布，能够方便地对齐布局内的组件。该布局和其他布局的组合可以实现更加丰富的布局方式。

第 3 行，指定 xml 命名空间。

第 4、5 行，指定布局组件的高度和宽度，可以有三种值。

1）float 类型，可以配置表示尺寸的 float 类型。可以是浮点数值，其默认单位为 px；也可以是带 px、vp、fp 单位的浮点数值；也可以引用 float 资源。

```
ohos:width="20"
ohos:width="20vp"
ohos:width="$float:size_value"
```

2）match_parent，表示控件宽度或高度与其父控件去掉内部边距后的宽度或高度相同。

3）match_content，表示控件宽度或高度由其包含的内容决定，包括其内容的宽度或高度以及内部边距的总和。

第 6 行，指定 DirectionalLayout 自身对齐方式，此处代码表示居中对齐。

第 7 行，指定 DirectionalLayout 布局方向，有两种值：vertical 和 horizontal。其中，"vertical" 布局方式如图 3-12 所示，DirectionalLayout 内部组件按垂直方向进行布局。

"horizontal" 布局方式如图 3-13 所示，DirectionalLayout 内部组件按水平方向进行布局。

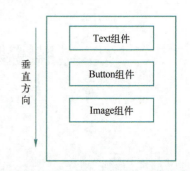

图 3-12　DirectionalLayout 垂直布局

图 3-13　DirectionalLayout 水平布局

第 9 行，一个界面布局会有很多组件构成，此处使用了一个文本组件 Text。

第 10 行，ohos:id，这个属性为该组件指定一个唯一的标识，在布局和 Ability 中可以通过这个 id 引用该组件。id 的格式如下。

```
ohos:id="$+id:text_helloworld"
```

id 前面有加号表示定义 id，如果没有加号，则表示引用已经定义的组件。

第 11、12 行，定义了组件的高度和宽度，有两个值可以选择。

1）match_parent，匹配父组件的宽度或高度。

2）match_content，根据自身内容大小将组件的宽度或高度调整到刚好能包裹自己。

第 13 行，定义了组件的背景元素，可以在 graphic 目录里创建 xml，然后在此处通过如下格式引用。

```
$graphic:graphic 目录里的文件名
```

第 14 行，定义了 DirectionalLayout 所包含的组件的对齐方式，各方式对应的值有如下几种。

1）left，左对齐。

2）top，顶部对齐。

3）right，右对齐。

4）bottom，底部对齐。

5）horizontal_center，水平居中。

6）vertical_center，垂直居中。

7）center，居中。

8）start，靠起始端对齐

9）end，靠结束端对齐

可以设置单个取值项，也可以使用"|"进行多项组合，示例如下。

```
ohos:layout_alignment="top"
ohos:layout_alignment="top|left"
```

第 15 行，指定 Text 组件要显示的内容，可以直接使用字符串，也可以引用 string 里的字符串定义值。

第 16 行，指定 Text 组件的文本大小。

要在 Splash 启动界面实现如图 3-14 所示界面，可以采用 DependentLayout 布局。

图 3-14　Splash 界面

完整 Splash 的布局的代码 ability_splash.xml 如下。

```
1.  <?xml version="1.0" encoding="utf-8"?>
2.  <DependentLayout
3.      xmlns:ohos="http://schemas.huawei.com/res/ohos"
```

```
4.      ohos:height="match_parent"
5.      ohos:width="match_parent"
6.      ohos:background_element="$graphic:background_ability_splash">
7.
8.      <Text
9.          ohos:id="$+id:txtSeconds"
10.         ohos:height="match_content"
11.         ohos:width="match_content"
12.         ohos:align_parent_right="true"
13.         ohos:text="6s"
14.         ohos:text_size="16fp"
15.         ohos:top_margin="5vp"
16.         ohos:right_margin="10vp"
17.         ohos:text_color="#EE0000"/>
18.     <Text
19.         ohos:id="$+id:txtCancel"
20.         ohos:height="match_content"
21.         ohos:width="match_content"
22.         ohos:left_of="$id:txtSeconds"
23.         ohos:right_margin="15vp"
24.         ohos:top_margin="5vp"
25.         ohos:text="$string:strCancel"
26.         ohos:text_size="16fp"
27.         ohos:text_color="#999"/>
28.     <Image
29.         ohos:height="150vp"
30.         ohos:width="150vp"
31.         ohos:image_src="$media:logo"
32.         ohos:horizontal_center="true"
33.         ohos:above="$id:txtTitle"
34.         ohos:scale_mode="inside"/>
35.     <Text
36.         ohos:id="$+id:txtTitle"
37.         ohos:height="match_content"
38.         ohos:width="match_content"
39.         ohos:text="$string:strSmartAgriculture"
40.         ohos:text_size="24fp"
41.         ohos:top_margin="30vp"
42.         ohos:center_in_parent="true"/>
43.     <Text
44.         ohos:id="$+id:txtPowered"
45.         ohos:height="match_content"
46.         ohos:width="match_content"
47.         ohos:above="$id:txtHarmonyOS"
48.         ohos:horizontal_center="true"
49.         ohos:text="Powered by"
50.         ohos:text_size="12fp"
```

```
51.            ohos:text_color="#888"/>
52.        <Text
53.            ohos:id="$+id:txtHarmonyOS"
54.            ohos:height="match_content"
55.            ohos:width="match_content"
56.            ohos:horizontal_center="true"
57.            ohos:align_parent_bottom="true"
58.            ohos:text="HarmonyOS"
59.            ohos:text_size="20fp"
60.            ohos:text_color="#888"
61.            ohos:bottom_margin="5vp"/>
62.
63.    </DependentLayout>
```

第 2 行，DependentLayout 布局是 Java UI 框架里的一种常见布局。与 DirectionalLayout 相比，拥有更多的排布方式，每个组件可以指定相对于其他同级元素的位置，或者指定相对于父组件的位置，如图 3-15 所示。

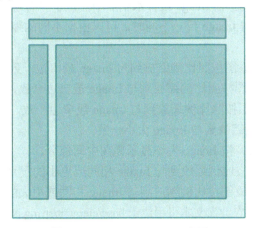

图 3-15 DependentLayout 布局

第 3～5 行，指定了 DependentLayout 的命名空间、宽和高。
第 6 行，引用 base/graphic/background_ability_splash.xml 文件自定义背景，该文件的代码如下。

```
<?xml version="1.0" encoding="UTF-8" ?>
<shape
    xmlns:ohos="http://schemas.huawei.com/res/ohos"
    ohos:shape="rectangle">

    <solid
        ohos:color="#EEF0F5"/>
</shape>
```

第 8～61 行为各组件相应代码，第 8～17 行，增加一个 Text 组件，存放 6s 倒计时，位于界面右上角。第 18～27 行，在 6s 倒计时组件左边增加一个"取消"Text 组件。第 28～34 行，在界面中心位置放置一个 Image 组件，用于存放 logo。第 35～42 行，在 logo 下方放置一个 Text 组件，存放"智慧农业"。第 43～61 行，在 Splash 界面底部放置两个 Text 组件，内容为"Powered by HarmonyOS"。

第 12 行，定义了倒计时位于父组件 DependentLayout 右边。
第 15 行，定义了倒计时文本上外边距为 5vp。
第 16 行，定义了倒计时文本右外边距为 10vp。
第 12、15、16 行确定了倒计时文本位于 Splash 界面右上角。
第 17 行，定义了文本的颜色。
第 22、23 行，定义了"取消"文本位于倒计时的左边。
第 25 行，引用了"取消"文本字符串，位于 base/element/string.json，增加的代码如下。

```
{
    "name": "strCancel",
    "value": "取消"
}
```

第 28 行，定义了一个 Image 组件，用于存放图片。
第 29、30 行，定义了图片组件的高和宽。
第 31 行，引用图片资源位于 base/media/logo.png。
第 32、33 行，分别定义了该 Image 组件水平居中，位于"智慧农业"文本的上方。
第 34 行，当图片尺寸与 Image 尺寸不同时，可以根据不同的缩放方式来对图片进行缩放。可以取如下值。

1）zoom_center：表示原图按照比例缩放到与 Image 最窄边一致，并居中显示。
2）zoom_start：表示原图按照比例缩放到与 Image 最窄边一致，并靠起始端显示。
3）zoom_end：表示原图按照比例缩放到与 Image 最窄边一致，并靠结束端显示。
4）stretch：表示将原图缩放到与 Image 大小一致。
5）center：表示不缩放，按 Image 大小显示原图中间部分。
6）inside：表示将原图按比例缩放到与 Image 相同或更小的尺寸，并居中显示。
7）clip_center：表示将原图按比例缩放到与 Image 相同或更大的尺寸，并居中显示。

第 39 行，定义了"智慧农业"文本引用位于 base/element/string.json 文件，增加的代码如下。

```
{
    "name": "strSmartAgriculture",
    "value": "智慧农业"
}
```

第 41、42 行，分别定义了"智慧农业"文本的上外边距，以及居中。
第 47、48 行，定义了 txtPowered 标识的组件位于 txtHarmonyOS 组件上方，并且居中。
第 49 行，定义了 ID 为 txtPowered 文本内容为"Powered by"。
第 56、57、61 行，定义了 ID 为 txtHarmonyOS 组件位于 Splash 中间，并且位于底部，距底部外边距 5vp。
第 58 行，定义了 ID 为 txtHarmonyOS 组件的文本为"HarmonyOS"。

3.4.6　编辑 Splash Ability

前面创建 Splash Ability 时创建了 SplashAbility 和 SplashAbilitySlice，希望能够使 Splash 界面停留 6s 后再进入登录页面（未登录时），如图 3-16 所示。

任务 3　创建 Splash 界面

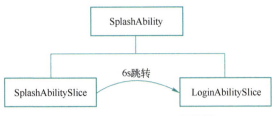

图 3-16　SplashAbility 逻辑图

　　SplashAbility 是 PageAbility，由两个 AbilitySlice 组成，分别为 SplashAbilitySlice 和 LoginAbilitySlice。SplashAbilitySlice 展现 App 启动欢迎界面逻辑，LoginAbilitySlice 展现登录界面逻辑。SplashAbilitySlice 在 6 秒倒计时之后会跳转到 LoginAbilitySlice。下面创建 LoginAbilitySlice。在 Ohos 项目视图下，右击 slice 目录，选择"New"→"Java Class"，如图 3-17 所示。

图 3-17　新建 Java Class

　　在弹出的"New Java Class"对话框中输入 LoginAbilitySlice，然后按〈Enter〉键，创建 LoginAbilitySlice 类，如图 3-18 所示。

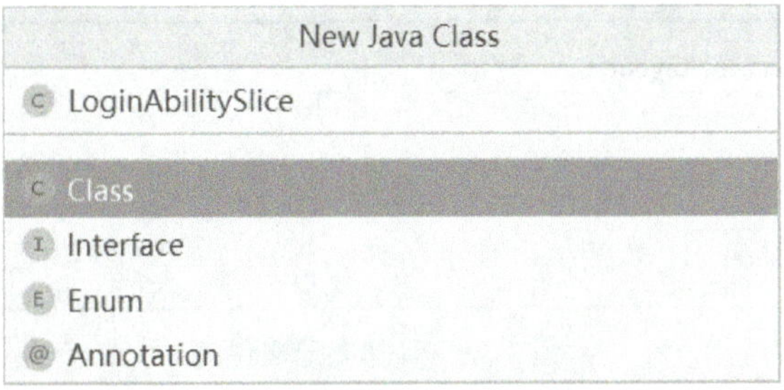

图 3-18　新建 Java 类

创建与 LoginAbilitySlice 绑定的布局文件，右击 layout 目录，选择"New"→"Layout Resource File"，如图 3-19 所示。

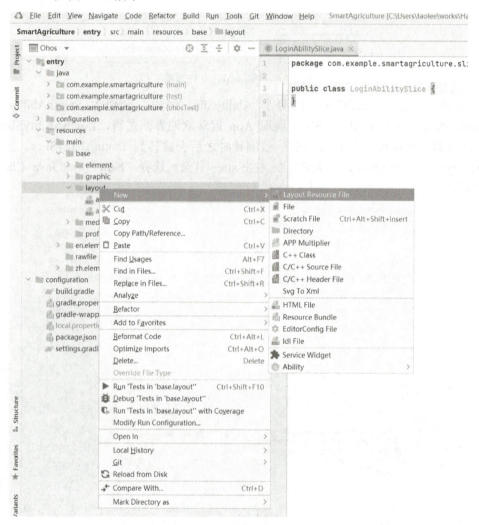

图 3-19　创建布局文件

输入 ability_login，然后单击"OK"按钮，创建 ability_login.xml 布局文件，如图 3-20 所示。

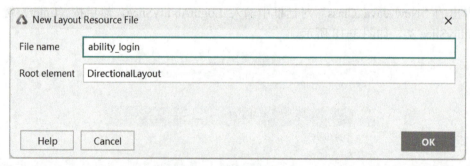

图 3-20　新建布局文件

在 ability_login.xml 文件中加入一个 Text 组件，用于临时调试 SplashAbilitySlice 跳转到

LoginAbilitySlice 时显示"Hello HarmonyOS",代码如下。

```
1.  <?xml version="1.0" encoding="utf-8"?>
2.  <DirectionalLayout
3.      xmlns:ohos="http://schemas.huawei.com/res/ohos"
4.      ohos:height="match_parent"
5.      ohos:width="match_parent"
6.      ohos:orientation="vertical">
7.
8.      <Text
9.          ohos:height="match_content"
10.         ohos:width="match_content"
11.         ohos:text="Hello HarmonyOS"
12.         ohos:text_size="18fp"/>
13. </DirectionalLayout>
```

通过如下代码,实现 LoginAbilitySlice 绑定 ability_login 布局,修改 LoginAbilitySlice.java 代码如下。

```
1.  package com.example.smartagriculture.slice;
2.
3.  import com.example.smartagriculture.ResourceTable;
4.  import ohos.aafwk.ability.AbilitySlice;
5.  import ohos.aafwk.content.Intent;
6.
7.  public class LoginAbilitySlice extends AbilitySlice {
8.      @Override
9.      protected void onStart(Intent intent) {
10.         super.onStart(intent);
11.         super.setUIContent(ResourceTable.Layout_ability_login);
12.     }
13. }
```

第 3 行,导入 ResourceTable,注意不是导入 ohos.global.systemres.ResourceTable。

第 9 行,重写 onStart 方法,该回调处于生命周期第一个方法。

第 11 行,将 ability_login 布局与 LoginAbilitySlice 绑定。其中以"ResourceTable.Layout_布局文件名"格式引用布局。

继续修改,完成 App 启动进入 Splash 欢迎界面,6s 之后进入 LoginAbilitySlice 页面。

```
1.  package com.example.smartagriculture.slice;
2.
3.  import com.example.smartagriculture.ResourceTable;
4.  import ohos.aafwk.ability.AbilitySlice;
5.  import ohos.aafwk.content.Intent;
6.
7.  import java.util.Timer;
8.  import java.util.TimerTask;
9.
10. public class SplashAbilitySlice extends AbilitySlice {
```

```
11.     private int seconds = 6;
12.     private Timer timer = null;
13.     @Override
14.     public void onStart(Intent intent) {
15.         super.onStart(intent);
16.         super.setUIContent(ResourceTable.Layout_ability_splash);
17.         timer = new Timer();
18.         TimerTask timerTask = new TimerTask() {
19.             @Override
20.             public void run() {
21.                 if(--seconds == 0) {
22.                     timer.cancel();
23.                     goToLogin();
24.                 }
25.             }
26.         };
27.         timer.schedule(timerTask, 0, 1000);
28.     }
29.
30.     private void goToLogin() {
31.         Intent intent = new Intent();
32.         present(new LoginAbilitySlice(), intent);
33.     }
34.
35.     @Override
36.     public void onActive() {
37.         super.onActive();
38.     }
39.
40.     @Override
41.     public void onForeground(Intent intent) {
42.         super.onForeground(intent);
43.     }
44. }
```

第 11、12 行，增加私有成员变量"秒 seconds"和"定时器 timer"。

第 17 行，定义定时器。

第 18~26 行，定义定时器任务，6s 倒计时后，删除定时器，调用 goToLogin 方法。

第 30~33 行，goToLogin 方法的实现。完成从 SplashAbilitySlice 跳转到 LoginAbilitySlice。在同一个 Page Ability 下面的 Silce 之间跳转，可以直接使用 present 方法进行跳转。在进行页面跳转时，用到了 Intent，Intent 是对象之间传递信息的载体。当一个 Ability 需要启动另一个 Ability 时，或者一个 AbilitySlice 需要导航到另一个 AbilitySlice 时，可以通过 Intent 指定启动的目标，同时携带相关数据。将 Intent 对象作为 present 的第二个参数，传递过去。present 的第一个参数指定了跳往的目的地的对象。

```
Intent intent = new Intent();
present(new TargetSlice(), intent);
```

第27行，启动定时器，schedule第一个参数指定任务，第二个参数指定延时执行的时间，单位为ms（毫秒），第三个参数指定执行周期，单位为ms（毫秒），此处指定1000ms，即1s。

如图3-21所示，完成SplashAbilitySlice跳转到LoginAbilitySlice。

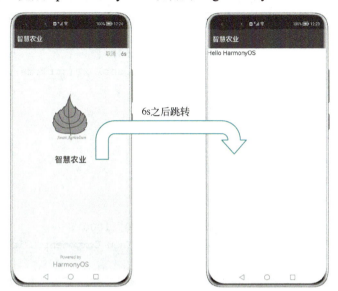

图3-21 SplashAbilitySlice跳转到LoginAbilitySlice

为了使SplashAbilitySlice页面的右上角倒计时动态减少，同时能够响应"取消"单击事件，即用户单击右上角的"取消"文本，则Splash欢迎界面直接进行跳转，需要更新SplashAbilitySlice.java，代码如下。

```
1.  package com.example.smartagriculture.slice;
2.
3.  import com.example.smartagriculture.ResourceTable;
4.  import ohos.aafwk.ability.AbilitySlice;
5.  import ohos.aafwk.content.Intent;
6.  import ohos.agp.components.Component;
7.  import ohos.agp.components.Text;
8.
9.  import java.util.Timer;
10. import java.util.TimerTask;
11.
12. public class SplashAbilitySlice extends AbilitySlice {
13.     private int seconds = 6;
14.     private Timer timer = null;
15.     private Text txtSeconds, txtCancel;
16.     @Override
17.     public void onStart(Intent intent) {
18.         super.onStart(intent);
19.         super.setUIContent(ResourceTable.Layout_ability_splash);
20.         txtSeconds = (Text)findComponentById(ResourceTable.Id_txtSeconds);
21.         txtCancel = (Text)findComponentById(ResourceTable.Id_txtCancel);
22.         timer = new Timer();
```

```java
23.         TimerTask timerTask = new TimerTask() {
24.             @Override
25.             public void run() {
26.                 String strTime = String.format("%d" + "s", seconds);
27.                 getUITaskDispatcher().asyncDispatch(new Runnable() {
28.                     @Override
29.                     public void run() {
30.                         txtSeconds.setText(strTime);
31.                     }
32.                 });
33.                 if(--seconds == 0) {
34.                     timer.cancel();
35.                     goToLogin();
36.                 }
37.             }
38.         };
39.         timer.schedule(timerTask, 0, 1000);
40.         txtCancel.setClickedListener(new Component.ClickedListener() {
41.             @Override
42.             public void onClick(Component component) {
43.                 timer.cancel();
44.                 goToLogin();
45.             }
46.         });
47.     }
48.
49.     private void goToLogin() {
50.         Intent intent = new Intent();
51.         present(new LoginAbilitySlice(), intent);
52.     }
53.
54.     @Override
55.     public void onActive() {
56.         super.onActive();
57.     }
58.
59.     @Override
60.     public void onForeground(Intent intent) {
61.         super.onForeground(intent);
62.     }
63. }
```

第 15 行，增加 txtSeconds 成员变量，用于存储界面右上角的倒计时组件；增加 txtCancel 成员变量，用于存储界面右上角的"取消"组件。

第 20 行，通过 findComponentById，传入 Text 组件的 id 标识符，赋值给 txtSeconds 变量来进行组件对象的获取。组件的引用需要的特定格式如下：

```
ResourceTable.type_name_组件的 id
```

其中 type_name 为类型名，如果是组件 id，则 type_name 为 id。组件的 ID 为定义在相关 AbilitySlice 的布局文件内的组件 id。

第 21 行，将界面右上角的"取消"组件复制给 txtCancel。

第 26 行，定义格式化字符串变量 strTime，格式为"数字+s"。

第 27~32 行，获取更新界面的 UI 线程，用于更新界面，此处将 strTime 字符串更新到 txtSeconds 所指定的组件界面里。注意，此处必须在 UI 线程中更新界面，如果直接在主线程更新，则会报如下错误。

```
java.lang.IllegalStateException: Attempt to update UI in non-UI thread.
```

第 40~46 行，设置组件"取消"的单击事件，当用户单击欢迎界面的"取消"时，系统自动运行这段代码，这段代码删除定时器，并且执行 goToLogin 方法，完成页面的直接跳转。

至此，Splash 的完整功能就完成了，如图 3-1 所示，6 秒倒计时到期自动跳转或者单击取消直接跳转页面。

3.5 提交代码到仓库

将当前工作目录下的文件加入暂存区，并提交到本地版本库，同时将任务 3 打上标签，具体流程如下。

```
1.  taolee@DESKTOP-0HJVM6A MINGW64~/works/HarmonyOS/Books/SmartAgriculture(master)
2.  $ git status
3.  On branch master
4.  Changes to be committed:
5.    (use "git restore --staged <file>..." to unstage)
6.          new file:   .idea/vcs.xml
7.
8.  Changes not staged for commit:
9.    (use "git add <file>..." to update what will be committed)
10.   (use "git restore <file>..." to discard changes in working directory)
11.         modified:   entry/src/main/config.json
12.         modified:   entry/src/main/resources/base/element/string.json
13.         modified:   entry/src/main/resources/en/element/string.json
14.         modified:   entry/src/main/resources/zh/element/string.json
15.
16. Untracked files:
17.   (use "git add <file>..." to include in what will be committed)
18.         .idea/previewer/
19.         entry/src/main/java/com/example/smartagriculture/SplashAbility.java
20.         entry/src/main/java/com/example/smartagriculture/slice/LoginAbilitySlice.java
21.         entry/src/main/java/com/example/smartagriculture/slice/SplashAbilitySlice.java
22.         entry/src/main/resources/base/graphic/background_ability_splash.xml
```

23.   entry/src/main/resources/base/layout/ability_login.xml
24.   entry/src/main/resources/base/layout/ability_splash.xml
25.
26.
27. taolee@DESKTOP-0HJVM6A MINGW64 ~/works/HarmonyOS/Books/SmartAgriculture (master)
28. $ git add .
29. warning:LF will be replaced by CRLF in.idea/previewer/phone/phoneSettingConfig_MateX2.json.
30. The file will have its original line endings in your working directory
31. warning:LF will be replaced by CRLF in.idea/previewer/phone/phoneSettingConfig_P40.json.
32. The file will have its original line endings in your working directory
33. warning:LF will be replaced by CRLF in.idea/previewer/previewConfigV2.json.
34. The file will have its original line endings in your working directory
35.
36. taolee@DESKTOP-0HJVM6A MINGW64 ~/works/HarmonyOS/Books/SmartAgriculture (master)
37. $ git commit -m "创建 Splash 界面"
38. [master 4a67167] 创建 Splash 界面
39. 14 files changed, 299 insertions(+), 3 deletions(-)
40. create mode 100644 .idea/previewer/phone/phoneSettingConfig_MateX2.json
41. create mode 100644 .idea/previewer/phone/phoneSettingConfig_P40.json
42. create mode 100644 .idea/previewer/previewConfigV2.json
43. create mode 100644 .idea/vcs.xml
44. create mode 100644 entry/src/main/java/com/example/smartagriculture/SplashAbility.java
45. create mode 100644 entry/src/main/java/com/example/smartagriculture/slice/LoginAbilitySlice.java
46. create mode 100644 entry/src/main/java/com/example/smartagriculture/slice/SplashAbilitySlice.java
47. create mode 100644 entry/src/main/resources/base/graphic/background_ability_splash.xml
48. create mode 100644 entry/src/main/resources/base/layout/ability_login.xml
49. create mode 100644 entry/src/main/resources/base/layout/ability_splash.xml
50.
51. taolee@DESKTOP-0HJVM6A MINGW64 ~/works/HarmonyOS/Books/SmartAgriculture (master)
52. $ git status
53. On branch master
54. nothing to commit, working tree clean
55.
56. taolee@DESKTOP-0HJVM6A MINGW64 ~/works/HarmonyOS/Books/SmartAgriculture (master)
57. $ git tag -a task3 -m "创建 Splash 界面"
58.
59. taolee@DESKTOP-0HJVM6A MINGW64 ~/works/HarmonyOS/Books/SmartAgriculture

```
(master)
    60. $ git log --pretty=oneline
    61. 4a671670a92413944ab77ba5c4f66fec96b6a2c7 (HEAD -> master, tag: task3)
```
创建 Splash 界面
```
    62. 4df17b7832e0746c1b154f22d34fea37069514c2 (tag: task1) 更改 App 图标和名称
    63. c8775743058113b200cb2e20a370e25a9056b365 Initial commit
```

第 2 行,通过 "git status" 查看当前工作区状态,可以看到有很多文件需要追踪。

第 28 行,通过 "git add ." 将当前目录下所有未跟踪的文件,加入跟踪。

第 37 行,将暂存区的内容提交到本地仓库,完成一次提交。

第 52 行,再次查看当前目录状态,可以看到目录干净。

第 57 行,通过 "git tag -a 标签名 -m 注释",将当前最新的提交打上标签。

第 60 行,打印日志,可以看到仓库最新版本处有标签 task3,即任务 3 的版本。

任务 4 创建新大陆云平台"智慧农业"项目

任务概述

本任务主要了解智慧农业项目的物联网体系结构,包括感知层与网络层相关技术。本书智慧农业项目的底层和云平台采用新大陆提供的技术方案,包括新大陆物联网云平台、新大陆物联网行业实训仿真软件、新大陆物联网云平台的数据模拟器和新大陆 1+X 传感网设备。智慧农业物联网体系如图 4-1 所示,HarmonyOS 应用与云平台进行交互,云平台的数据可以来自三个路径,包括虚拟仿真,实物设备和数据模拟器,其中虚拟仿真即新大陆物联网行业实训仿真软件,实物设备即新大陆 1+X 传感网设备,数据模拟器即新大陆物联网云平台的数据模拟器。读者只需选择新大陆物联网云平台三个数据来源中的一种技术方案,将数据上传到云平台即可。

图 4-1 智慧农业物联网系统

知识目标

- 了解新大陆物联网云平台。
- 了解新大陆物联网云平台的数据模拟器。
- 了解新大陆物联网行业实训仿真软件。
- 了解新大陆 1+X 传感网设备体系。

技能目标

- 能使用仿真软件搭建感知层环境。
- 能创建新大陆物联网云平台项目。
- 能使用新大陆物联网云平台的数据模拟器产生测试数据。
- 能使用新大陆 1+X 传感网设备搭建感知层环境。

4.1 创建云平台项目

智慧农业不是一个单独的 App,而是基于物联网的智慧农业 App。智慧农业 App 需要和云平台进行交互,此处云平台采用了新大陆物联网云平台。

4.1.1 了解新大陆物联网云平台

新大陆物联网云平台是基于智能传感器、无线传输技术、大规模数据处理与远程控制等物联网核心技术与互联网、无线通信、云计算、大数据技术高度融合开发的一套物联网云服务平台，集设备在线采集、远程控制、无线传输、数据分析、预警信息发布、决策支持、一体化控制等功能于一体的物联网系统。用户及管理人员可以通过手机、平板计算机、计算机等信息终端，实时掌握传感设备数据，及时获取报警、预警信息，并可以手动或自动的调整控制设备，最终实现使以上管理变得轻松简单的目标。同时，新大陆物联网云平台也是针对物联网教育、科研推出的开放的物联网云服务教学平台，可作为中职、高职和本科的物联网应用技术（物联网工程）专业课的主要实训设备。

4.1.2 创建新大陆物联网云平台"智慧草坪"项目

1. 登陆云平台

登录网址 http://www.nlecloud.com，如果没有账号，注册账号，然后登录，单击右上角的"账号（手机号）"→"开发设置"，确认 ApiKey 有没有过期，如果过期，则重新生成 ApiKey，如图 4-2 所示。

图 4-2　生成 ApiKey

2. 创建项目

（1）新增项目

单击"开发者中心"，然后单击"新增项目"，在弹出的"添加项目"对话框中，可对"项目名称""行业类别""联网方案"等信息进行填充。

在本案例中，设置"项目名称"为"智慧农业"，"行业类别"选择"智慧农业"，其他保持默认，如图 4-3 所示。

图 4-3 云平台新建项目

（2）添加设备

项目新建完毕，可为其添加设备。需要对"设备名称""通信协议""设备标识"进行设置（设备标识不可重复），如图 4-4 所示。

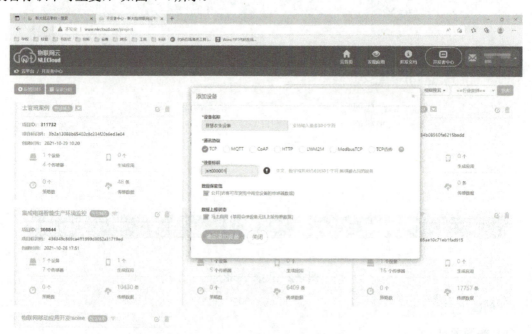

图 4-4 云平台添加设备

单击"确定添加设备"，添加设备完成后如图 4-5 所示，并将图 4-4 中的"设备标识"和"传输密钥"记下，配置网关时需要用到这些信息。

图 4-5　添加设备完成效果

4.1.3　创建传感器

如图 4-6 所示，在传感器子页面单击 "+" 可以创建传感器。

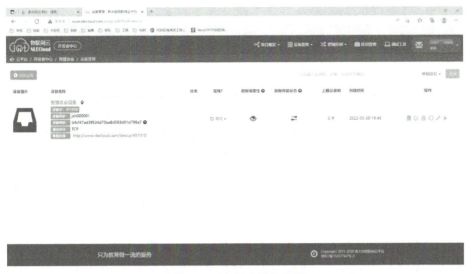

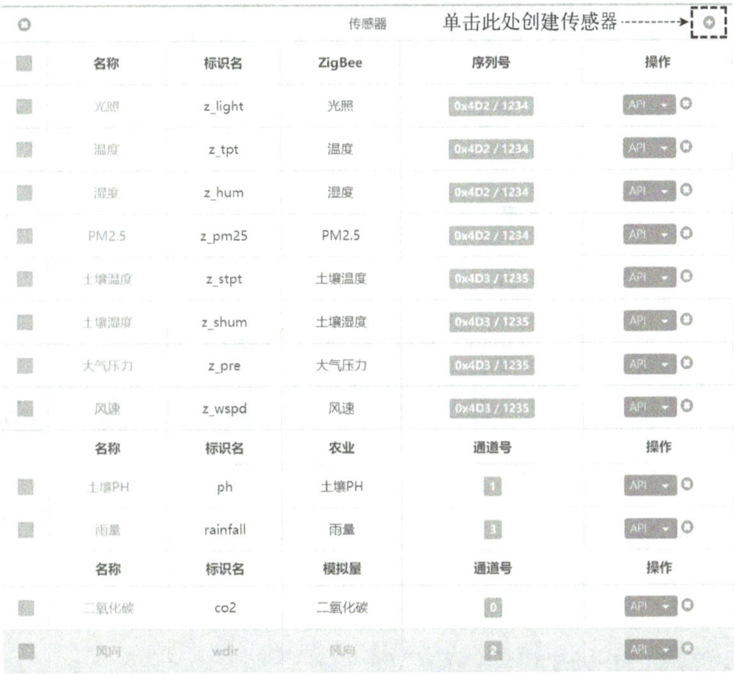

图 4-6　创建传感器

4.1.4　创建执行器

如图 4-7 所示，在执行器子页面单击 "+" 可以创建执行器。

图 4-7 创建执行器

4.2 创建物联网行业实训仿真项目

新大陆物联网行业实训仿真平台是一款虚拟的物联网系统安装与维护的学习平台,它有着高真实度的实验设备与实验过程,模拟与实际操作高度贴合的实验平台,给学生、老师身临其境之感,能够覆盖现阶段物联网教学中常用的设备,旨在向学员提供一种在仿真实验环境中练习物联网感知层的设备搭建环境。

4.2.1 创建仿真项目

结合本书"智慧农业"实例,创建如图 4-8 所示的仿真项目。

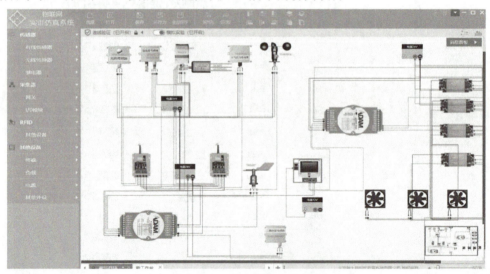

图 4-8 仿真项目

4.2.2 调试智慧农业数据采集和控制

在仿真平台,单击模拟实验,可以上传数据到云平台,能够快速获取系统各种数据信息,如图 4-9 所示。
1)大气环境中的温度值、湿度值、风速、风向、光照、气压、PM2.5 及 CO_2 含量。
2)土壤环境的 pH 值、雨量、温湿度等。
3)执行器控制模块,实现了对风扇的控制。

图 4-9　仿真数据上传云平台

4.3 使用新大陆物联网云平台数据模拟器

把新大陆物联网云平台数据模拟器作为云平台数据来源中的一种，是开发测试 App 最方便的一种方法，可以屏蔽系统底层的技术细节，专心于 App 应用层相关功能的实现。下面学习数据模拟器的使用。

单击"调试工具"→"数据模拟"→"启动虚拟设备"，可以使用云平台模拟数据，如图 4-10 所示。

图 4-10　启动虚拟设备

启动虚拟设备后，可以模拟设备上报数据，如图 4-11 所示。

图 4-11　模拟设备上报数据

4.4 使用新大陆 1+X 传感网设备设计原型

在实际的智慧农业数据流交互中，云平台的数据应该来自真实的设备产生的数据。新大陆物联网云平台支持新大陆 1+X 传感网设备数据的上传，所以此处可以直接采用新大陆 1+X 传感网设备设计物联网的感知层。

4.4.1 认识新大陆 1+X 传感网设备体系

新大陆 1+X 传感网设备以模块化的方式提供最小单片机系统，如图 4-12 所示。所有通信系统都是由单片机加通信模组构成。整个体系包括了有线传输和无线传输。

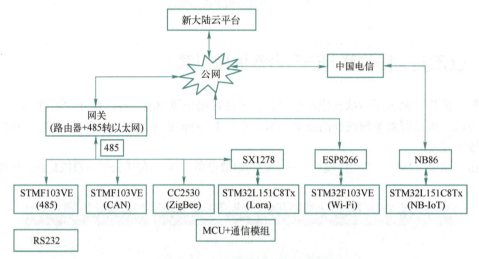

图 4-12 新大陆 1+X 传感网设备体系

1. 有线方式

RS-232、RS-485 和 CAN 通过有线方式将数据传递到网关，网关内部将数据重新封帧，组成以太网包，进而传递到云平台。

2. 无线方式

ZigBee、LoRa 构成短距无线通信，各 ZigBee 之间、各 LoRa 之间通过无线采集数据，最终都是通过有线方式，将数据上传到网关，后续步骤和有线方式一样。

Wi-Fi 方式，是直接将数据路由到局域网，进而将数据上传到云平台。

NB-IoT 方式，通过与周围的基站交互，将数据传递到公网，进而将数据上传到云平台。

4.4.2 搭建感知层原型

采集温湿度数据和火焰数据的例子如图 4-13 所示，温湿度、火焰数据通过 ZigBee 传递到汇聚节点，汇聚节点通过串口转换接口，将数据以 485 方式上传到网关，网关内部进行数据重新封帧，变成以太网包并传递到云平台，进而完成感知层数据的上传。

任务 4　创建新大陆云平台"智慧农业"项目

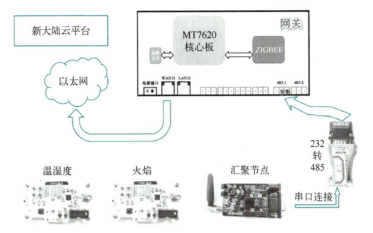

图 4-13　1+X 设备示例

任务 5 RESTful API 调试

任务概述

本任务主要学习 HTTP 的基础、掌握 HTTP 的概念、请求报文的格式、响应报文的格式、常见的 HTTP 请求方法以及响应代码的分类和含义。然后，学习 REST 的架构，了解 RESTful 的概念，了解其 URL 的设计规则。最后学习 Postman 软件，使用 Postman 软件调试新大陆物联网云平台的 RESTful API 接口。

知识目标

- 掌握 HTTP 基本概念。
- 掌握 HTTP 响应报文的格式。
- 掌握 HTTP 常见的响应代码。
- 掌握 Postman 调试 API 结构。
- 掌握 HTTP 请求报文的格式。
- 掌握 HTTP 常见请求方法。
- 了解 RESTful 架构的含义。
- 掌握新大陆物联网云平台核心 API 的接口。

技能目标

- 能使用 Postman 调试 Web API 接口。
- 能分析 URL 的基本结构。

5.1 HTTP 基础

HTTP（HyperText Transfer Protocol，超文本传输协议）是因特网上应用最为广泛的一种网络传输协议。它是一个简单的请求-响应协议，通常运行在 TCP 之上。它指定了客户端可能发送给服务器什么样的消息以及得到什么样的响应。

5.1.1 HTTP 消息结构

HTTP 是基于客户端/服务端（C/S）的架构模型，通过一个可靠的链接来交换信息，是一个无状态的请求-响应协议。HTTP 客户端一般是 Web 浏览器，连接到服务器，向服务器发起请求，服务器返回响应。

一个 HTTP 请求包含请求行、请求头部、空行和请求数据，如图 5-1 所示。

图 5-1 HTTP 请求报文格式

一个 HTTP 的响应包含响应状态行、响应报头、空行、响应数据，如图 5-2 所示。

图 5-2　HTTP 响应报文格式

以访问百度为例，代码如下。

客户端请求：

1. GET / HTTP/1.1
2. User-Agent: PostmanRuntime/7.28.4
3. Accept: */*
4. Postman-Token: 4f883cf9-1edf-4db7-a4c4-7d0e3ca07a23
5. Host: www.baidu.com
6. Accept-Encoding: gzip, deflate, br
7. Connection: keep-alive

服务端响应：

1. HTTP/1.1 200 OK
2. Accept-Ranges: bytes
3. Cache-Control: no-cache
4. Connection: keep-alive
5. Content-Length: 227
6. Content-Type: text/html
7. Date: Wed, 22 Sep 2021 09:31:48 GMT
8. P3p: CP=" OTI DSP COR IVA OUR IND COM "
9. P3p: CP=" OTI DSP COR IVA OUR IND COM "
10. Pragma: no-cache
11. Server: BWS/1.1
12. Set-Cookie: BD_NOT_HTTPS=1; path=/; Max-Age=300
13. Set-Cookie: BIDUPSID=56F3DDA0D62AFBD8AF1FA32EA3EF66B0; expires=Thu, 31-Dec-37 23:55:55 GMT; max-age=2147483647; path=/; domain=.baidu.com
14. Set-Cookie: PSTM=1632303108; expires=Thu, 31-Dec-37 23:55:55 GMT;max-age=2147483647; path=/; domain=.baidu.com
15. Set-Cookie: BAIDUID=56F3DDA0D62AFBD8B597DE1E24034CD9:FG=1; max-age=31536000; expires=Thu, 22-Sep-22 09:31:48 GMT; domain=.baidu.com; path=/; version=1; comment=bd
16. Strict-Transport-Security: max-age=0
17. Traceid: 1632303108035040820218202561155568791401
18. X-Frame-Options: sameorigin
19. X-Ua-Compatible: IE=Edge,chrome=1
20.

```
21. <html>
22. <head>
23. <script>
24. location.replace(location.href.replace("https://","http://"));
25. </script>
26. </head>
27. <body>
28. <noscript><meta http-equiv="refresh" content="0;url=http://www.baidu.com/"></noscript>
29. </body>
30. </html>
```

5.1.2 HTTP 方法

HTTP1.0 定义了三种请求方法：GET、POST 和 HEAD。

HTTP1.1 新增了六种请求方法：OPTIONS、PUT、PATCH、DELETE、TRACE 和 CONNECT。具体说明如表 5-1 所示。

表 5-1 HTTP 方法

方法	描述
GET	请求指定的页面信息
POST	传输实体主体，数据被包含在请求体中
HEAD	获取报文头部
OPTIONS	询问支持的方法
PUT	传输文件
PATCH	是对 PUT 方法的补充，用来对已知资源进行局部更新
DELETE	删除文件
TRACE	回显服务器收到的请求，主要用于测试或诊断
CONNECT	要求用隧道协议连接代理

5.1.3 HTTP 常用方法 GET 和 POST

1. GET

GET 提交的数据会直接填充在请求报文的 URL 上，如"https://www.baidu.com/s?ie=UTF-8&wd=HarmonyOS%20%E9%B8%BF%E8%92%99"。其中，"?"划分域名和 GET 提交的参数，ie=UTF-8 中的 ie 是参数名，UTF-8 是参数值，多个参数之间用"&"进行分割，如果参数值是中文，则会转换成诸如"%20%E9%B8%BF%E8%92%99"。一般来说，浏览器处理的 URL 最大长度为 1024B（不同浏览器有差异），所以 GET 方法提交参数长度有限制。

2. POST

POST 方法提交的数据会附在正文上，一般请求正文的长度是没有限制的，但表单中所能处理的长度一般为 100KB（不同浏览器有差异），且考虑报文的传输效率，不推荐过长。

所以 GET 方法可以用来传输一些可以公开、不敏感的参数信息，而 POST 方法可以用来提交一个用户的敏感信息（如果不使用 HTTPS 加密，报文正文仍旧是明文，容易被截获读取，所以建议使用 HTTPS）。

5.1.4　HTTP 常见请求头部

1．Accept

告知服务端，客户端接收什么类型的响应。

2．Referer

表示该请求是从哪个 URL 进来的。

3．Cache-Control

对缓存进行控制。

4．Accept-Encoding

告知服务器能接受什么编码格式，包括字符编码及压缩形式（一般都是压缩形式）。

5．Host

指定要请求的资源所在的主机和端口。

6．User-Agent

告知服务器，客户端的操作系统、浏览器的版本和名称等。

5.1.5　HTTP 常见响应报头

1．Cache-Control

服务端通过该属性告诉客户端该怎么控制响应内容的缓存。

2．ETag

表示请求资源的版本，如果该资源发生变化，那么这个属性也会跟着变。

3．Location

在重定向中或者创建新资源时使用。

4．Set-Cookie

服务端可以设置客户端的 cookie。

5．Allow

服务器支持哪些请求方法（如 GET、POST 等）。

6．Content-Encoding

文档的编码（Encode）方法。只有在解码之后才可以得到 Content-Type 头指定的内容类型。利用 gzip 压缩文档能够显著地减少 HTML 文档的下载时间。

7．Content-Length

表示内容长度。只有当浏览器使用 HTTP 持久连接时才需要这个数据。

8．Content-Type

文档的类型。

5.1.6 HTTP 状态码

在服务端的响应报文状态行中有状态码,服务器返回一个包含 HTTP 状态码的信息头用以响应浏览器的请求。常见的状态码如下。

1)200 表示请求成功。
2)301 表示资源被永久转移到其他 URL。
3)303 表示重定向,把请求重定向到其他页面。
4)404 表示请求的资源不存在。
5)500 表示服务器内部错误。

HTTP 状态码由三个十进制数字组成,第一个十进制数字定义了状态码的类型,后两个数字没有分类的作用。HTTP 状态码共分为 5 种类型,如表 5-2 所示。

表 5-2 HTTP 状态码类别

状态码	说明
100~199	成功接收请求,需要请求者继续执行操作
200~299	成功完成整个处理请求
300~399	重定向,需要进一步的操作以完成请求
400~499	客户端请求错误
500~599	服务器端出现错误

5.2 RESTful 架构

一个 Web 服务遵循 REST 规范将会获得如下特性。
1)URL 具有很强可读性且具有自描述性。
2)资源描述与视图的松耦合。
3)可提供 RESTful API,便于第三方系统集成,提高互操作性等。

5.2.1 REST 概述

REST 全称是 Representational State Transfer,是描述性状态迁移。Roy Fielding 博士提出了 REST 这一软件架构,目的是在符合该架构原理的前提下,理解和评估以网络为基础的应用软件的架构设计,得到一个功能强、性能好、适宜通信的架构。REST 指的是一组架构约束条件和原则,如果一个架构符合 REST 的约束条件和原则,我们就称它为 RESTful 架构。

5.2.2 资源与 URI

REST 中的"描述"指的是资源。在 Web 里,任何事物,只要有被引用到的必要,它就是一个资源。资源可以是实体(例如账号),也可以只是一个抽象概念(例如价值)。而统一资源标识符(Uniform Resource Identifier,URI),一个用于标识某一互联网资源名称的字符串,表示的就是 Web 上所有可用的资源,如 HTML 文档、图像、视频片段、程序等都是由 URI 进行标识的。该种标识允许用户对任何(包括本地和互联网)资源都能通过特定的协议进行交互操作。

URL 是 URI 的一个子集。它是 Uniform Resource Locator 的缩写，译为"统一资源定位符"。它是 URI 的表现形式之一。URL 的格式由下列三部分组成。

第一部分是协议（或称为服务方式）；第二部分是存有该资源的主机 IP 地址（有时也包括端口号）；第三部分是主机资源的具体地址。

URI 的具体格式如下。

[协议名]://[用户名]:[密码]@[服务器地址]:[服务器端口号]/[路径]?[查询字符串]#[片段ID]

1）登录信息（用户名：密码）：表示服务器认证的账号。此项是可选项。
2）服务器地址：表示地址，可以是域名，如 www.HarmonyOS.com，或者是 IPv4、IPv6 地址。
3）服务器端口号：指定服务器连接的网络端口号。常见的是 80，也是默认端口号，可以省略。此选项是可选项。
4）路径：表示服务器上的文件路径。
5）查询字符串：表示使用查询字符串传入任意参数。此选项是可选项。
6）片段 ID：表示标记出已获取资源中的子资源（文档内的某个位置）。此选项是可选项。
举例如下。

https://developer.harmonyos.com/cn/docs/documentation/doc-guides/document-outline-0000001064589184，其中，协议名：https，服务器地址：developer.harmonyos.com，路径：cn/docs/documentation/doc-guides/document-outline-0000001064589184

5.2.3 统一资源接口

RESTful 架构应该遵循统一接口原则，使用标准的 HTTP 方法如 GET、PUT 和 POST 对资源进行操作，并遵循这些方法的语义。URI 表示资源的名称，一般是名词，而不应该包括资源的操作。下面这些例子是不符合要求的设计。

1）GET /getID/1
2）POST /createAccount
3）DELETE /deleteAccount/1

5.3 使用 Postman 调试 API 接口

Postman 是一个用于构建和使用 API 的 API 平台。Postman 简化了 API 生命周期的每个步骤并简化了协作，因此开发人员可以更快地创建更好的 API。该平台包含一套全面的工具，可帮助加速 API 生命周期——从设计、测试、文档和模拟到 API 的共享和可发现性。

开发人员可以用 Postman 进行网页调试与发送网页 HTTP 请求，模拟 GET 或者 POST 或者其他方式的请求来调试接口。

5.3.1 Postman 安装

访问 https://www.postman.com/，下载并安装 Postman App。

5.3.2 Postman 基本使用

Postman 创建请求如图 5-3 所示,单击"+"创建请求。

图 5-3 Postman 创建请求

如图 5-4 所示,①选择 HTTP 方法,此处选择 GET,②输入请求地址,③单击"Send",发起请求。可以看到底部区域显示了服务器返回的响应内容。在该例子中,也设置了请求的参数,除了可以设置请求参数外,也可以设置认证信息,Header 头部信息,Body 请求内容等。

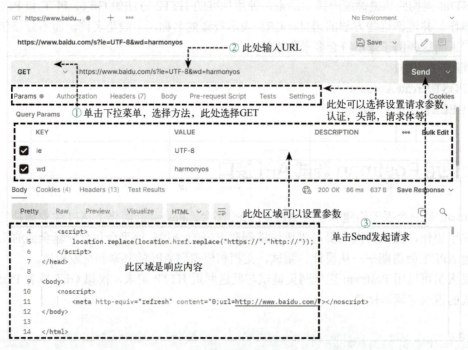

图 5-4 请求设置

如图 5-5 所示,单击"Console",可以在控制台查看请求响应的细节。

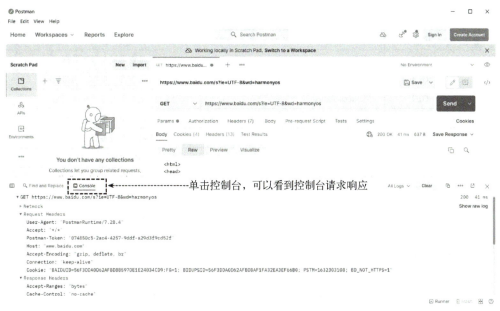

图 5-5 控制台查看

5.4 调试新大陆物联网云平台 API 接口

新大陆物联网云平台提供了 RESTful API，方便第三方系统接入，也方便 HarmonyOS App 的接入。为了后续开发 App 与云平台交互，接下来要先学习新大陆物联网云平台提供的 RESTful API。

5.4.1 归纳新大陆物联网云平台 RESTful API

访问新大陆物联网云平台（http://www.nlecloud.com/）官网，可以查看平台的开发者模块，里面有 RESTful API 文档。所有接口的分类如表 5-3 所示。

表 5-3 新大陆物联网云平台 RESTful API

组 别	HTTP	API 地址	描 述
账号 API	POST	users/login	用户登录（同时返回 AccessToken）
	GET	users/apikey	查询用户 APIKey
	PUT	users/apikey	更新用户 APIKey
	POST	users/login	用户登录（同时返回 AccessToken）
项目 API	GET	projects/{projectid}	查询单个项目
	GET	projects	模糊查询项目
	POST	projects	新增项目
	PUT	projects/{projectid}	更新项目
	DELETE	projects	删除项目
	GET	projects/{projectid}/sensors	查询项目下所有设备的传感器

(续)

组 别	HTTP	API 地址	描 述
设备 API	GET	devices/datas	批量查询设备最新数据
	GET	devices/simulationdatas	批量查询设备实时模拟数据
	GET	devices/status	批量查询设备的在线状态
	GET	devices/{deviceid}	查询单个设备
	GET	devices	模糊查询设备
	POST	devices	添加新设备
	PUT	devices/{deviceid}	更新某个设备
	DELETE	devices/{deviceid}	删除某个设备
设备传感器 API	GET	devices/{deviceid}/sensors/{apitag}	查询单个传感器
	GET	devices/{deviceid}/sensors	模糊查询传感器
	POST	devices/{deviceid}/sensors	添加新传感器
	PUT	devices/{deviceid}/sensors/{apitag}	更新某个传感器
	DELETE	devices/{deviceid}/sensors/{apitag}	删除某个传感器
传感数据 API	GET	devices/{deviceid}/datas/grouping	聚合查询传感数据
	GET	devices/{deviceid}/datas	模糊查询传感数据
	POST	devices/{deviceid}/datas	上传传感数据
	GET	bindata/{fileid}	获取文件
策略 API	GET	strategys	查询单个策略
	GET	strategys	查询策略
	POST	strategys	新增策略
	DELETE	strategys	删除策略
	GET	strategys/records	查询策略执行记录
	PUT	strategys/{strategyid}	更新策略
	POST	strategys/enable/{strategyid}	启用/禁用策略
命令 API	POST	cmds	发送命令/控制设备

5.4.2 调试用户登录 API

1. 请求方式及地址

```
POST
http://api.nlecloud.com/Users/Login
```

2. 包体请求参数

包体请求参数如表 5-4 所示。

表 5-4 请求参数

参数	类型	描述	其他
Account	string	用户名	Required
Password	string	用户密码	RequiredData type: PasswordString length: inclusive between 0 and 32
IsRememberMe	boolean	记住密码	

3. 请求示例

```
{
  "Account": "sample string 1",
  "Password": "sample string 2",
  "IsRememberMe": true
}
```

4. 响应参数

响应参数详情如表 5-5 所示。

表 5-5 响应参数

参 数	类 型	描 述	其 他
ResultObj	AccountLoginResultDTO		
Status	ResultStatus	Status	
StatusCode	integer	返回的状态码	
Msg	string	返回的消息	
ErrorObj	Object		

5. 示例

如图 5-6 所示，按照序号顺序操作。①选择 HTTP 请求方式为 POST。②输入用户登录的 URL。③选择配置请求体 Body。④选择 raw。⑤选择 JSON 格式。⑥输入 JSON 格式的账号、密码。⑦单击 Send，发送请求。可以在下方看到服务的响应，其中有用户的基本信息。

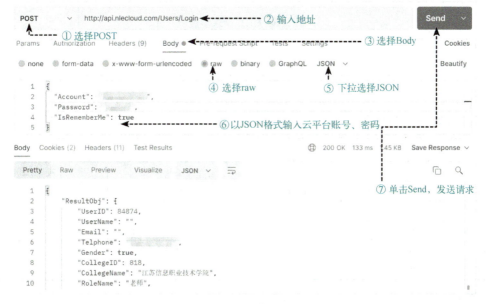

图 5-6 用户登录

如图 5-7 所示，用户登录成功。服务器返回的数据中，最重要的数据是"AccessToken"，用户与服务器后续的交互操作，必须携带该值。

图 5-7　AccessToken

5.4.3　查询设备最新数据

1. 请求方式及地址

GET
http://api.nlecloud.com/Devices/{devIds}

2. URL 请求参数

URL 请求参数如表 5-6 所示。

表 5-6　URL 请求参数

参　数	类　型	描　述	其　他
devIds	string	设备 ID 用逗号隔开，限制 100 个设备	Required

3. 响应参数

响应参数如表 5-7 所示。

表 5-7　响应参数

参　数	类　型	描　述	其　他
ResultObj	Collection of DeviceSensorDataDTO		
Status	ResultStatus	Status	
StatusCode	Integer	返回的状态码	
Msg	String	返回的消息	
ErrorObj	Object	ErrorObj	

4. 响应示例

如图 5-8 所示，按照序号顺序操作。①选择 HTTP 请求方法 GET。②输入请求 URL。③选择 Headers，以配置请求头部。④增加 AccessToken，其值为用户登录接口返回的 AccessToken 值。⑤单击发送请求。

任务 5 RESTful API 调试

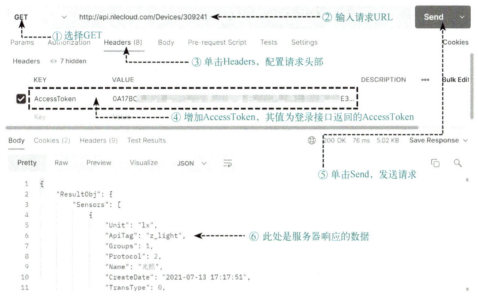

图 5-8 请求设备最新数据

以下是返回的智慧农业的传感器的数据示例，格式为 JSON。

```
1.    {
2.        "ResultObj": {
3.            "Sensors": [
4.                {
5.                    "Unit": "lx",
6.                    "ApiTag": "z_light",
7.                    "Groups": 1,
8.                    "Protocol": 2,
9.                    "Name": "光照",
10.                   "CreateDate": "2021-07-13 17:17:51",
11.                   "TransType": 0,
12.                   "DataType": 1,
13.                   "TypeAttrs": "",
14.                   "DeviceID": 309241,
15.                   "SensorType": "light",
16.                   "GroupID": null,
17.                   "Coordinate": null,
18.                   "Value": 10000.0,
19.                   "RecordTime": "2021-09-09 08:59:50"
20.                },
21.                {
22.                    "Unit": "℃",
23.                    "ApiTag": "z_tpt",
24.                    "Groups": 1,
25.                    "Protocol": 2,
26.                    "Name": "温度",
27.                    "CreateDate": "2021-07-13 17:19:06",
28.                    "TransType": 0,
```

```
29.                    "DataType": 1,
30.                    "TypeAttrs": "",
31.                    "DeviceID": 309241,
32.                    "SensorType": "temperature",
33.                    "GroupID": null,
34.                    "Coordinate": null,
35.                    "Value": 15.59,
36.                    "RecordTime": "2021-09-09 08:59:50"
37.                },
38.                {
39.                    "Unit": "%RH",
40.                    "ApiTag": "z_hum",
41.                    "Groups": 1,
42.                    "Protocol": 2,
43.                    "Name": "湿度",
44.                    "CreateDate": "2021-07-13 17:19:28",
45.                    "TransType": 0,
46.                    "DataType": 1,
47.                    "TypeAttrs": "",
48.                    "DeviceID": 309241,
49.                    "SensorType": "humidity",
50.                    "GroupID": null,
51.                    "Coordinate": null,
52.                    "Value": 50.0,
53.                    "RecordTime": "2021-09-09 08:59:50"
54.                },
55.                {
56.                    "Unit": "μg/m3",
57.                    "ApiTag": "z_pm25",
58.                    "Groups": 1,
59.                    "Protocol": 2,
60.                    "Name": "PM2.5",
61.                    "CreateDate": "2021-07-13 17:39:36",
62.                    "TransType": 0,
63.                    "DataType": 1,
64.                    "TypeAttrs": "",
65.                    "DeviceID": 309241,
66.                    "SensorType": "pm2p5",
67.                    "GroupID": null,
68.                    "Coordinate": null,
69.                    "Value": 150.0,
70.                    "RecordTime": "2021-09-09 08:59:50"
71.                },
72.                {
73.                    "Unit": "℃",
74.                    "ApiTag": "z_stpt",
75.                    "Groups": 1,
```

```
76.                "Protocol": 2,
77.                "Name": "土壤温度",
78.                "CreateDate": "2021-07-13 17:40:37",
79.                "TransType": 0,
80.                "DataType": 1,
81.                "TypeAttrs": "",
82.                "DeviceID": 309241,
83.                "SensorType": "soil-temperature",
84.                "GroupID": null,
85.                "Coordinate": null,
86.                "Value": 20.0,
87.                "RecordTime": "2021-09-09 08:59:50"
88.            },
89.            {
90.                "Unit": "%RH",
91.                "ApiTag": "z_shum",
92.                "Groups": 1,
93.                "Protocol": 2,
94.                "Name": "土壤湿度",
95.                "CreateDate": "2021-07-13 17:40:58",
96.                "TransType": 0,
97.                "DataType": 1,
98.                "TypeAttrs": "",
99.                "DeviceID": 309241,
100.                "SensorType": "soil-humidity",
101.                "GroupID": null,
102.                "Coordinate": null,
103.                "Value": 50.0,
104.                "RecordTime": "2021-09-09 08:59:50"
105.            },
106.            {
107.                "Unit": "kPa",
108.                "ApiTag": "z_pre",
109.                "Groups": 1,
110.                "Protocol": 2,
111.                "Name": "大气压力",
112.                "CreateDate": "2021-07-13 17:42:39",
113.                "TransType": 0,
114.                "DataType": 1,
115.                "TypeAttrs": "",
116.                "DeviceID": 309241,
117.                "SensorType": "pressure",
118.                "GroupID": null,
119.                "Coordinate": null,
120.                "Value": 55.0,
121.                "RecordTime": "2021-09-09 08:59:50"
122.            },
```

```
123.                {
124.                    "Unit": "m/s",
125.                    "ApiTag": "z_wspd",
126.                    "Groups": 1,
127.                    "Protocol": 2,
128.                    "Name": "风速",
129.                    "CreateDate": "2021-07-13 17:42:53",
130.                    "TransType": 0,
131.                    "DataType": 1,
132.                    "TypeAttrs": "",
133.                    "DeviceID": 309241,
134.                    "SensorType": "wind-speed",
135.                    "GroupID": null,
136.                    "Coordinate": null,
137.                    "Value": 15.0,
138.                    "RecordTime": "2021-09-09 08:59:50"
139.                },
140.                {
141.                    "Unit": "pH",
142.                    "ApiTag": "ph",
143.                    "Groups": 1,
144.                    "Protocol": 1,
145.                    "Name": "土壤pH",
146.                    "CreateDate": "2021-07-13 17:47:02",
147.                    "TransType": 0,
148.                    "DataType": 1,
149.                    "TypeAttrs": "",
150.                    "DeviceID": 309241,
151.                    "SensorType": "soilph",
152.                    "GroupID": null,
153.                    "Coordinate": null,
154.                    "Value": 2.0,
155.                    "RecordTime": "2021-09-09 08:59:41"
156.                },
157.                {
158.                    "Unit": "mm/min",
159.                    "ApiTag": "rainfall",
160.                    "Groups": 1,
161.                    "Protocol": 1,
162.                    "Name": "雨量",
163.                    "CreateDate": "2021-07-15 22:47:52",
164.                    "TransType": 0,
165.                    "DataType": 1,
166.                    "TypeAttrs": "",
167.                    "DeviceID": 309241,
168.                    "SensorType": "rainfall",
169.                    "GroupID": null,
```

```
170.                    "Coordinate": null,
171.                    "Value": 23.0,
172.                    "RecordTime": "2021-09-09 08:59:41"
173.                },
174.                {
175.                    "Unit": "ppm",
176.                    "ApiTag": "m_co2",
177.                    "Groups": 1,
178.                    "Protocol": 1,
179.                    "Name": "CO₂",
180.                    "CreateDate": "2021-09-06 14:12:11",
181.                    "TransType": 0,
182.                    "DataType": 1,
183.                    "TypeAttrs": "",
184.                    "DeviceID": 309241,
185.                    "SensorType": "co2",
186.                    "GroupID": null,
187.                    "Coordinate": null,
188.                    "Value": 2500.0,
189.                    "RecordTime": "2021-09-09 08:59:50"
190.                },
191.                {
192.                    "Unit": "°",
193.                    "ApiTag": "m_wind_direction",
194.                    "Groups": 1,
195.                    "Protocol": 1,
196.                    "Name": "风向",
197.                    "CreateDate": "2021-09-06 14:12:25",
198.                    "TransType": 0,
199.                    "DataType": 1,
200.                    "TypeAttrs": "",
201.                    "DeviceID": 309241,
202.                    "SensorType": "wind-direction",
203.                    "GroupID": null,
204.                    "Coordinate": null,
205.                    "Value": 180.0,
206.                    "RecordTime": "2021-09-09 08:59:50"
207.                },
208.                {
209.                    "OperType": 1,
210.                    "OperTypeAttrs": "",
211.                    "ApiTag": "m_fan2",
212.                    "Groups": 2,
213.                    "Protocol": 1,
214.                    "Name": "风扇",
215.                    "CreateDate": "2021-07-13 17:48:45",
216.                    "TransType": 1,
```

```
217.                    "DataType": 2,
218.                    "TypeAttrs": "",
219.                    "DeviceID": 309241,
220.                    "SensorType": "fan",
221.                    "GroupID": null,
222.                    "Coordinate": null,
223.                    "Value": false,
224.                    "RecordTime": "2021-09-09 08:59:50"
225.                },
226.                {
227.                    "OperType": 1,
228.                    "OperTypeAttrs": "",
229.                    "ApiTag": "m_fan3",
230.                    "Groups": 2,
231.                    "Protocol": 1,
232.                    "Name": "风扇",
233.                    "CreateDate": "2021-07-13 17:48:56",
234.                    "TransType": 1,
235.                    "DataType": 2,
236.                    "TypeAttrs": "",
237.                    "DeviceID": 309241,
238.                    "SensorType": "fan",
239.                    "GroupID": null,
240.                    "Coordinate": null,
241.                    "Value": false,
242.                    "RecordTime": "2021-09-09 08:59:50"
243.                },
244.                {
245.                    "OperType": 1,
246.                    "OperTypeAttrs": "",
247.                    "ApiTag": "m_fan4",
248.                    "Groups": 2,
249.                    "Protocol": 1,
250.                    "Name": "风扇",
251.                    "CreateDate": "2021-07-13 17:49:07",
252.                    "TransType": 1,
253.                    "DataType": 2,
254.                    "TypeAttrs": "",
255.                    "DeviceID": 309241,
256.                    "SensorType": "fan",
257.                    "GroupID": null,
258.                    "Coordinate": null,
259.                    "Value": false,
260.                    "RecordTime": "2021-09-09 08:59:50"
261.                },
262.                {
263.                    "OperType": 1,
```

```
264.                    "OperTypeAttrs": "",
265.                    "ApiTag": "m_fan1",
266.                    "Groups": 2,
267.                    "Protocol": 1,
268.                    "Name": "风扇",
269.                    "CreateDate": "2021-09-06 11:25:51",
270.                    "TransType": 1,
271.                    "DataType": 2,
272.                    "TypeAttrs": "",
273.                    "DeviceID": 309241,
274.                    "SensorType": "fan",
275.                    "GroupID": null,
276.                    "Coordinate": null,
277.                    "Value": false,
278.                    "RecordTime": "2021-09-09 08:59:50"
279.                }
280.            ],
281.            "DeviceID": 309241,
282.            "Name": "物联网网关",
283.            "Tag": "P97E1525354",
284.            "SecurityKey": "228318e69dde45c78a9cb656ea4692f6",
285.            "ProjectID": 278986,
286.            "Protocol": "TCP",
287.            "IsOnline": false,
288.            "LastOnlineIP": "121.235.142.165",
289.            "LastOnlineTime": "2021-09-09 08:59:41",
290.            "Coordinate": "",
291.            "CreateDate": "2021-07-13 17:15:04",
292.            "IsShare": true,
293.            "IsTrans": true
294.        },
295.        "Status": 0,
296.        "StatusCode": 0,
297.        "Msg": null,
298.        "ErrorObj": null
299.    }
```

5.4.4 模糊查询传感器

1. 请求方式及地址

GET
http://api.nlecloud.com/devices/{deviceId}/Sensors

2. URL 请求参数

URL 请求参数如表 5-8 所示。

表 5-8 请求参数

参　数	类　型	描　述
deviceId	integer	设备 ID（必须）
apiTags	string	传感标识名（必须），多个标识名之间用逗号分开（参数值为空时查询所有传感器）

3. 响应参数

响应参数如表 5-9 所示。

表 5-9 响应参数

参　数	类　型	描　述	其　他
ResultObj	Collection of SensorBaseInfoDTO		
Status	ResultStatus	Status	
StatusCode	Integer	返回的状态码	
Msg	String	返回的消息	
ErrorObj	Object		

4. 响应示例

如图 5-9 所示。①选择 HTTP 方法 GET。②输入请求 URL。③配置 URL 参数。

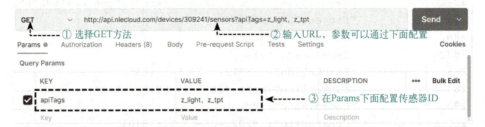

图 5-9 配置请求方法、参数

如图 5-10 所示。④配置 Headers，为请求头部增加 AccessToken。⑤单击 Send，发送请求。

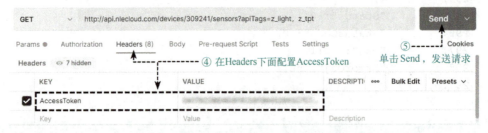

图 5-10 配置 AccessToken

以下是服务器的响应数据。

```
1.  {
2.      "ResultObj": [
3.          {
4.              "Unit": "lx",
5.              "ApiTag": "z_light",
6.              "Groups": 1,
```

```
7.            "Protocol": 2,
8.            "Name": "光照",
9.            "CreateDate": "2021-07-13 17:17:51",
10.           "TransType": 0,
11.           "DataType": 1,
12.           "TypeAttrs": "",
13.           "DeviceID": 309241,
14.           "SensorType": "light",
15.           "GroupID": null,
16.           "Coordinate": null,
17.           "Value": 10000.0,
18.           "RecordTime": "2021-09-09 08:59:50"
19.       },
20.       {
21.           "Unit": "℃",
22.           "ApiTag": "z_tpt",
23.           "Groups": 1,
24.           "Protocol": 2,
25.           "Name": "温度",
26.           "CreateDate": "2021-07-13 17:19:06",
27.           "TransType": 0,
28.           "DataType": 1,
29.           "TypeAttrs": "",
30.           "DeviceID": 309241,
31.           "SensorType": "temperature",
32.           "GroupID": null,
33.           "Coordinate": null,
34.           "Value": 15.59,
35.           "RecordTime": "2021-09-09 08:59:50"
36.       }
37.   ],
38.   "Status": 0,
39.   "StatusCode": 0,
40.   "Msg": null,
41.   "ErrorObj": null
42. }
```

5.4.5 发送命令控制设备

1. 请求方式及地址

```
Post
http://api.nlecloud.com/cmds
```

2. URL 请求参数

URL 请求参数如表 5-10 所示。

表 5-10 请求参数

参　数	类　型	描　述	其　他
deviceId	integer	设备 ID（必填）	Required
apiTag	string	传感标识名（参数必须有但值可为空）	Required

3. HTTP Body 内容

用户自定义数据：integer/float/json/string/二进制等类型值，示例如下。

开关类：开 = 1，关 = 0，暂停 = 2
家居类：调光灯亮度=0～254，RGB 灯色度=2～239，窗帘、卷闸门、幕布打开百分比=3%～100%，红外指令=1(on) 或 2(off)

4. 响应参数

响应参数如表 5-11 所示。

表 5-11 响应参数

参　数	类　型	描　述	其　他
ResultObj	String	命令 ID，GUID 格式的字符串，平台范围内唯一	
Status	ResultStatus		
StatusCode	Integer	返回的状态码	
Msg	String	返回的消息	
ErrorObj	Object		

5. 响应示例

如图 5-11 所示。①选择 HTTP 方法 POST。②输入请求 URL。③配置设备参数。

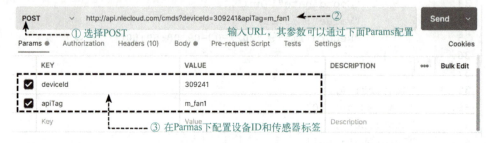

图 5-11 配置发送命令的参数

如图 5-12 所示。④配置 Headers，请求头部增加 AccessToken。

图 5-12 配置 AccessToken

如图 5-13 所示。⑤配置 Body 请求体，选择"Body"。⑥选择"raw"。⑦选择"JSON"。

⑧输入"1",开启设备。⑨单击"Send"发送请求。

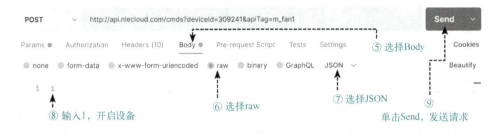

5-13 配置 Body

以下是服务器的响应数据。

1. {
2.   "ResultObj": "09296b2c-b817-4bda-bc22-b9d0bcc8672c",
3.   "Status": 0,
4.   "StatusCode": 0,
5.   "Msg": null,
6.   "ErrorObj": null
7. }

如果设备没有上线,则返回如下数据。

1. {
2.   "ResultObj": "231bd844-df83-4f2b-bb73-178f1ddd5ed0",
3.   "Status": 1,
4.   "StatusCode": 0,
5.   "Msg": "可能设备 ID【309241】还未上线!\nMaybe the device ID [309241] is not online yet!",
6.   "ErrorObj": null
7. }

任务 6　创建登录功能

任务概述

本任务主要完成登录功能的创建。如图 6-1 所示，登录界面提供用户账号、密码以及登录按钮，可以让用户输入云平台账号，登录到云平台。App 根据云平台提供的 RESTful API 接口，登录云平台，获得 AccessToken（访问令牌），以便后续的任务可以和云平台进行正常业务交互。

图 6-1　登录界面

知识目标

- 掌握 Image 组件、TextField 组件、Button 组件。
- 掌握布局和常用组件的基本属性。
- 掌握获取网络权限的方式。
- 掌握蒹葭（JianJia）网络库。

技能目标

- 能使用 TextField 组件进行设计。
- 能使用蒹葭（JianJia）网络库访问云平台。
- 能使用 Button 组件完成单击事件的处理。

6.1　编辑登录界面 ability_login.xml

登录界面在之前的任务中已经创建，现在修改登录界面，使其具有如图 6-1 所示效果。ability_login.xml 文件的代码如下。

```
1.  <?xml version="1.0" encoding="utf-8"?>
2.  <DependentLayout
```

```
3.          xmlns:ohos="http://schemas.huawei.com/res/ohos"
4.          ohos:height="match_parent"
5.          ohos:width="match_parent"
6.          ohos:background_element="$graphic:background_ability_login">
7.
8.      <Image
9.          ohos:id="$+id:imgLogo"
10.         ohos:height="120vp"
11.         ohos:width="120vp"
12.         ohos:image_src="$media:logo"
13.         ohos:horizontal_center="true"
14.         ohos:top_margin="200vp"
15.         ohos:scale_mode="inside"/>
16.
17.     <TextField
18.         ohos:id="$+id:tfAccount"
19.         ohos:height="50vp"
20.         ohos:width="match_parent"
21.         ohos:left_margin="20vp"
22.         ohos:right_margin="20vp"
23.         ohos:left_padding="15vp"
24.         ohos:hint="$string:strTfTelephone"
25.         ohos:hint_color="#FF355343"
26.         ohos:text_input_type="pattern_number"
27.         ohos:text_alignment="vertical_center"
28.         ohos:below="$id:imgLogo"
29.         ohos:text_size="28fp"
30.         ohos:top_margin="50vp"
31.         ohos:background_element="$graphic:login_textfield_bg"/>
32.     <TextField
33.         ohos:id="$+id:tfPasswd"
34.         ohos:height="50vp"
35.         ohos:width="match_parent"
36.         ohos:left_margin="20vp"
37.         ohos:right_margin="20vp"
38.         ohos:left_padding="15vp"
39.         ohos:hint="$string:strTfPassword"
40.         ohos:hint_color="#FF355343"
41.         ohos:text_alignment="vertical_center"
42.         ohos:below="$id:tfAccount"
43.         ohos:text_size="28fp"
44.         ohos:top_margin="20vp"
45.         ohos:text_input_type="pattern_password"
46.         ohos:background_element="$graphic:login_textfield_bg"/>
47.
48.     <Button
49.         ohos:id="$+id:btnLogin"
50.         ohos:height="50vp"
```

```
51.            ohos:width="275vp"
52.            ohos:text="$string:strBtnLogin"
53.            ohos:text_size="20fp"
54.            ohos:text_alignment="center"
55.            ohos:text_weight="800"
56.            ohos:align_parent_bottom="true"
57.            ohos:bottom_margin="80vp"
58.            ohos:horizontal_center="true"
59.            ohos:background_element="$graphic:login_btn_bg"
60.            />
61. </DependentLayout>
```

1. 设置页面背景

第 6 行，设置页面背景，引用 background_ability_login.xml，在 base/graphic 下创建该文件。background_ability_login.xml 完整代码如下。

```
1. <?xml version="1.0" encoding="utf-8"?>
2. <shape
3.     xmlns:ohos="http://schemas.huawei.com/res/ohos"
4.     ohos:shape="rectangle">
5.
6.     <solid
7.         ohos:color="#EEF0F5"/>
8. </shape>
```

2. 展示 Logo

第 8~15 行使用了 Image 组件展示 Logo，Image 组件继承自 Component 组件，Image 的常用属性如表 6-1 所示。

表 6-1 Image 常用属性

属性名称	中文描述	取值	取值说明	使用案例
image_src	图像	Element 类型	可直接配置色值，也可引用 color 资源或引用 media/graphic 下的图片资源	ohos:image_src="#FFFFFFFF" ohos:image_src="$color:black" ohos:image_src="$media:warning" ohos:image_src="$graphic:graphic_src"
scale_mode	图像缩放类型	zoom_center	表示原图按照比例缩放到与 Image 最窄边一致，并居中显示	ohos:scale_mode="center"
		zoom_start	表示原图按照比例缩放到与 Image 最窄边一致，并靠起始端显示	
		zoom_end	表示原图按照比例缩放到与 Image 最窄边一致，并靠结束端显示	
		stretch	表示将原图缩放到与 Image 大小一致	
		center	表示不缩放，按 Image 大小显示原图中间部分	
		inside	表示将原图按比例缩放到与 Image 相同或更小的尺寸，并居中显示	
		clip_center	表示将原图按比例缩放到与 Image 相同或更大的尺寸，并居中显示	

第 12 行，设置图片源属性，位于 base/media/logo.png。

第 13 行，设置图片水平居中。

第 14 行，设置图片距离上外边距 200vp。

第 15 行，设置图片的缩放模式为 inside，即将图片按比例缩放至与 Image 组件相同尺寸，且居中显示。

3. 用户账号输入

第 17~31 行增加了 TextField 组件，用于输入账号，此处账号是用户注册新大陆物联网云平台的手机号。

TextField 提供了一种文本输入框，用户输入账号和密码后，获取用户输入的内容并提交给服务器进行认证。TextField 继承自 Text，Text 组件的常用的属性如表 6-2 所示。

表 6-2 Text 常用属性

属性名称	中文描述	取值	取值说明	使用案例
text	显示文本	string 类型	可以直接设置文本字串，也可以引用 string 资源（推荐使用）	ohos:text="熄屏时间" ohos:text="$string:test_str"
hint	提示文本	string 类型	可以直接设置文本字串，也可以引用 string 资源（推荐使用）	ohos:hint="联系人" ohos:hint="$string:test_str"
text_size	文本大小	float 类型	表示字体大小的 float 类型。可以是浮点数值，其默认单位为 px；也可以是带 px、vp、fp 单位的浮点数值；也可以引用 float 资源	ohos:text_size="30" ohos:text_size="16fp" ohos:text_size="$float:size_value"
text_color	文本颜色	color 类型	可以直接设置色值，也可以引用 color 资源	ohos:text_color="#A8FFFFFF" ohos:text_color="$color:black"
hint_color	提示文本颜色	color 类型	可以直接设置色值，也可以引用 color 资源	ohos:hint_color="#A8FFFFFF" ohos:hint_color="$color:black"
text_input_type	文本输入类型	pattern_null	表示未指定文本输入类型，默认文本输入类型为内容模式	ohos:text_input_type="pattern_null"
		pattern_text	表示文本输入类型为普通文本模式	ohos:text_input_type="pattern_text"
		pattern_number	表示文本输入类型为数字	ohos:text_input_type="pattern_number"
		pattern_password	表示文本输入类型为密码	ohos:text_input_type="pattern_password"

第 21~23 行，设置 TextField 组件的左右外边距，以及左填充距大小。

第 24 行，设置 TextField 的提示内容，此处使用字符串引用方式，引用内容在 base/element/string.json 文件里，具体如下：

```
{
    "name": "txtField_telephone",
    "value": "请输入手机号"
}
```

第 25 行，设置提示文本的颜色。

第 26 行，设置 TextField 输入内容的类型，此处设置为 pattern_number，即必须输入数字。

第 27 行，设置文本的对齐方式为 vertical_center，垂直居中。

第 28 行，设置账号位于 logo 下方。

第 29 行，设置文本字体的大小，此处为 28fp，其中"fp"是字体大小单位。

第 30 行，设置组件上外边距。

第 31 行，设置背景，文件位于 base/graphic/login_textfield_bg.xml，可以通过右击"graphic"目录，然后在弹出的菜单中选择"New"→"Graphic Resource File"进行创建，完整代码如下：

```
1.  <?xml version="1.0" encoding="UTF-8" ?>
2.  <shape
3.      xmlns:ohos="http://schemas.huawei.com/res/ohos"
```

```
4.        ohos:shape="rectangle">
5.
6.        <corners ohos:radius="20"/>
7.        <stroke ohos:color="#FF0E0E10"/>
8.
9.        <solid
10.           ohos:color="#FF80A2F8"/>
11. </shape>
```

该段代码的含义如下。

第 4 行，定义了背景形状为 rectangle，即矩形。

第 6 行，设置了矩形的圆角弧度半径为 20。

第 7 行，设置了线条颜色为#FF0E0E10。

第 9 行，设置了矩形的背景色为#FF80A2F8。

4．用户密码输入

第 32～46 行实现用户密码输入，也是使用 TextField 组件，密码的输入与用户账号的输入大同小异。

第 36～38 行，设置密码输入的左右外边距，以及左填充大小。

第 39 行，设置提示文本，使用字符串引用的方式，位于 base/element/string.json，具体如下。

```
{
    "name": "strTfPassword",
    "value": "请输入密码"
}
```

第 40 行，设置提示文本的颜色。

第 41 行，设置文本垂直居中。

第 42 行，设置密码位于账号的下方，此处直接引用账号的 id，格式是：$id:组件 id。

第 43 行，设置文本字体的大小为 28fp。

第 44 行，设置密码的上外边距为 20vp，其中"vp"是长度度量单位。

第 45 行，设置 TextField 输入的类型，此处值为 pattern_password，即输入的内容是密码，此时组件会在用户输入内容时自动采用密文代替，如小圆点。

第 46 行，设置密码的背景，此处设置和账号样式保持一致，即引用文件 base/graphic/login_textfield_bg.xml。

5．登录按钮

第 48～60 行使用 Button 组件完成登录按钮设计。该组件可以提供与用户单击交互的事件处理。Button 组件是一种常见的组件，单击可以触发对应的操作，通常由文本或图标组成，也可以由图标和文本共同组成。Button 组件没有自有的 XML 属性，共有 XML 属性继承自 Text 组件。

第 52 行，设置 Button 显示的文本，内容位于文件 base/element/string.json 里的字符串 btn_login，具体如下。

```
{
    "name": "strBtnLogin",
    "value": "登录->"
}
```

第 53 行，设置 Button 按钮的文本的大小为 20fp。

第 54 行，设置文本的对齐方式为 center，即居中。

第 55 行，设置文本的粗细，此处为 800。

第 56 行，设置 Button 组件位于父布局的底部，此处的父布局是 DependentLayout，即位于手机页面底部。

第 57 行，设置 Button 组件下外边距为 80vp。

第 58 行，设置 Button 组件水平居中。

第 59 行，设置 Button 的背景，创建 login_btn_bg.xml，位于 base/graphic。login_btn_bg.xml 完整代码如下。

```
1.  <?xml version="1.0" encoding="UTF-8" ?>
2.  <state-container xmlns:ohos="http://schemas.huawei.com/res/ohos">
3.      <item ohos:state="component_state_pressed" ohos:element="$graphic:login_btn_bg_pressed"/>
4.      <item ohos:state="component_state_empty" ohos:element="$graphic:login_btn_bg_empty"/>
5.  </state-container>
```

该段代码的含义如下。

第 2 行，使用了 state-container 元素，其中设置了两种状态，一个是按钮被按下时的状态，一个是按钮未按下时的状态。

第 3 行，设置 Button 按钮按下时，其状态引用 base/graphic/login_btn_bg_pressed.xml，该文件需要在 base/graphic 目录里创建，设置背景色为#FF8986EE，login_btn_bg_pressed.xml 的完整代码如下。

```
1.  <?xml version="1.0" encoding="UTF-8" ?>
2.  <shape xmlns:ohos="http://schemas.huawei.com/res/ohos"
3.         ohos:shape="rectangle">
4.      <solid
5.         ohos:color="#FF8986EE"/>
6.      <corners
7.         ohos:radius="30"/>
8.  </shape>
```

第 4 行，设置按钮未按下时的状态，其状态引用 login_btn_bg_empty.xml，该文件同样需要在 base/graphic 目录里创建，设置背景色为#FF33DDEC，login_btn_bg_empty.xml 的完整代码如下。

```
1.  <?xml version="1.0" encoding="UTF-8" ?>
2.  <shape xmlns:ohos="http://schemas.huawei.com/res/ohos"
3.         ohos:shape="rectangle">
4.      <solid
5.         ohos:color="#FF33DDEC"/>
6.      <corners
7.         ohos:radius="30"/>
8.  </shape>
```

6.2 编辑登录逻辑 LoginAbilitySlice.java

LoginAbilitySlice.java 文件在任务 4 里已经创建，现在修改其代码，使用户单击登录按钮时，在 DevEco Studio Log 窗口打印用户输入的账号和密码，代码如下。

```
1.  package com.example.smartagriculture.slice;
2.
3.  import com.example.smartagriculture.ResourceTable;
4.  import ohos.aafwk.ability.AbilitySlice;
5.  import ohos.aafwk.content.Intent;
6.  import ohos.agp.components.Button;
7.  import ohos.agp.components.Component;
8.  import ohos.agp.components.TextField;
9.
10. public class LoginAbilitySlice extends AbilitySlice {
11.
12.     private Button btnLogin;
13.     private TextField tfAccount, tfPasswd;
14.     @Override
15.     protected void onStart(Intent intent) {
16.         super.onStart(intent);
17.         super.setUIContent(ResourceTable.Layout_ability_login);
18.         btnLogin = (Button) findComponentById(ResourceTable.Id_btnLogin);
19.         tfAccount = (TextField) findComponentById(ResourceTable.Id_tfAccount);
20.         tfPasswd = (TextField) findComponentById(ResourceTable.Id_tfPasswd);
21.
22.         btnLogin.setClickedListener(new Component.ClickedListener() {
23.             @Override
24.             public void onClick(Component component) {
25.                 login(tfAccount.getText().trim(), tfPasswd.getText().trim());
26.             }
27.         });
28.     }
29.
30.     private void login(String account, String passwd) {
31.         System.out.println("账号：" + account + "\n" + "密码：" + passwd);
32.     }
33. }
```

第 22～27 行，设置按钮的单击事件处理。设置按钮单击事件处理可以通过匿名内部类实现，结构如下。

```
new Component.ClickedListener() {
    @Override
    public void onClick(Component component) {
        //具体单击事件处理逻辑
    }
});
```

当单击按钮的时候，会进入 onClick 方法处理流程，在 onClick 方法里调用了 login 方法，传递界面的账号和密码。getTxet()获取界面的字符串内容，然后调用 trim()移除字符串两侧的空白字符。

第 30 行，login 方法的具体实现，使用 System.out.println 打印账号和密码。

编译运行代码，效果如图 6-2 所示，单击账号输入框，在弹出的虚拟键盘里输入账号和密码。

图 6-2　虚拟键盘输入界面

如图 6-3 所示，按照序号顺序操作。① 单击"DevEco Studio"底部的"Log"，② 指定运行的 App 包名。通过滚动日志区域，可以看到打印的日志内容。在日志内容过多的情况下，还可以通过过滤日志来查找目标日志。

图 6-3　查看日志

将此部分功能保存到本地版本库，版本的日志为"实现登录页面读取账号和密码，并打印日志"。

6.3 引入网络库兼葭（JianJia）

兼葭（JianJia）是一款 HarmonyOS 上的网络请求处理框架，它源自安卓的 Retrofit 框架。兼葭不仅能实现 Retrofit 的功能，还会提供一些 Retrofit 没有的功能。例如，国内的应用一般都是有多个域名的，Retrofit 不支持动态替换域名，而兼葭支持动态域名替换。

6.3.1 添加 mavenCentral() 仓库

在项目根目录下的 build.gradle 文件中添加 mavenCentral()仓库，以 Ohos 项目视图为例（若本书不做特别说明，则项目视图默认为 Ohos 视图），路径为 configuration/build.gradle，代码如下。

```
1.  buildscript {
2.      repositories {
3.          mavenCentral()
4.          maven {
5.              url 'https://repo.huaweicloud.com/repository/maven/'
6.          }
7.          maven {
8.              url 'https://developer.huawei.com/repo/'
9.          }
10.     }
11.     dependencies {
12.         classpath 'com.huawei.ohos:hap:3.0.5.2'
13.         classpath 'com.huawei.ohos:decctest:1.2.7.2'
14.     }
15. }
16.
17. allprojects {
18.     repositories {
19.         mavenCentral()
20.         maven {
21.             url 'https://repo.huaweicloud.com/repository/maven/'
22.         }
23.         maven {
24.             url 'https://developer.huawei.com/repo/'
25.         }
26.     }
27. }
```

第 3、19 行，添加 mavenCentral 中央仓库，然后单击"Sync Now"同步配置。

6.3.2 添加依赖

在 entry 目录下的 build.gradle 文件中的 dependencies 闭包下添加下面的依赖。路径为 entry/configuration/build.gradle，代码如下。

任务 6 创建登录功能

```
1.  dependencies {
2.      implementation fileTree(dir: 'libs', include: ['*.jar', '*.har'])
3.      testImplementation 'junit:junit:4.13.1'
4.      ohosTestImplementation 'com.huawei.ohos.testkit:runner:2.0.0.200'
5.      // 蒹葭的核心代码
6.      implementation 'io.gitee.zhongte:jianjia:1.0.1'
7.      // 数据转换器，数据转换器使用 gson 来帮我们解析 json，不需要我们手动解析 json
8.      implementation 'io.gitee.zhongte:converter-gson:1.0.1'
9.      implementation "com.google.code.gson:gson:2.8.2"
10.     // 日志拦截器，通过日志拦截器可以看到请求头、请求体、响应头、响应体
11.     implementation 'com.squareup.okhttp3:logging-interceptor:3.7.0'
12.     // 如果服务端返回的 json 有特殊字符，比如中文的双引号，gson 在解析的时候会对特殊字符进行转义
13.     // 这时就需要将转义后的字符串进行反转义，commons-lang 可以对特殊字符进行转义和反转义
14.     implementation 'commons-lang:commons-lang:2.6'
15. }
```

第 5~14 行，增加蒹葭依赖，其中以//开头的为注释行。

6.3.3 增加网络权限和 HTTP 访问

在配置文件 config.json 中添加 HTTP 访问功能，代码如下。

```
1.  "deviceConfig": {
2.      "default": {
3.        "network": {"cleartextTraffic": true}
4.      }
5.  },
```

第 2~4 行，增加了 HTTP 访问功能，如果不配置 cleartextTraffic，则必须使用 HTTPS 访问，本 App 与新大陆物联网云平台交互时使用的协议为 HTTP，所以此处必须加上该配置。

增加网络访问权限，代码如下。

```
1.  "distro": {
2.      "deliveryWithInstall": true,
3.      "moduleName": "entry",
4.      "moduleType": "entry",
5.      "installationFree": false
6.  },
7.  "reqPermissions": [{"name": "ohos.permission.INTERNET"}],
8.  "abilities": [
9.      {
```

第 7 行，增加了网络访问权限，App 联网访问时必须申请网络权限。

6.4 登录云平台

使用蒹葭网络库，连接新大陆物联网云平台，登录并获取 AccessToken，为后续的网络请求

提供认证。

6.4.1 创建 Wan 接口

在 com.example.smartlawn.net 下创建 Wan 接口，把所有的请求放在一个接口里面即可，没必要创建多个接口类。Wan 完整代码如下。

```
1.   package com.example.smartagriculture.net;
2.   import com.example.smartagriculture.bean.Account;
3.   import poetry.jianjia.Call;
4.   import poetry.jianjia.http.*;
5.
6.   public interface Wan {
7.       @POST("users/login")
8.       @FormUrlEncoded
9.       Call<Account> login(@Field("Account") String account, @Field("Password") String password,
10.                          @Field("IsRememberMe") boolean isRememberMe);
11.  }
```

第 9 行，增加 login 方法。

6.4.2 创建 Account Bean

在任务 5 中，使用 Postman 访问新大陆物联网云平台 RESTful API 时，可以获得返回内容，其中登录返回示例如下（其中 AccessToken 是我们最关心的内容，一些隐私信息使用了"xxxx"代替）。

```
1.   {
2.       "ResultObj": {
3.           "UserID": 84874,
4.           "UserName": "",
5.           "Email": "",
6.           "Telephone": "xxxx",
7.           "Gender": true,
8.           "CollegeID": 818,
9.           "CollegeName": "江苏信息职业技术学院",
10.          "RoleName": "老师",
11.          "RoleID": 8,
12.          "AccessToken": "xxxx",
13.          "AccessTokenErrCode": 0,
14.          "ReturnUrl": null,
15.          "DataToken": "9e10059ed1dc4b1f"
16.      },
17.      "Status": 0,
18.      "StatusCode": 0,
19.      "Msg": null,
20.      "ErrorObj": null
```

```
21. }
```

在 com.example.smartlawn.bean 下创建与登录响应内容相匹配的 Account 类，代码如下。

```
1.  package com.example.smartagriculture.bean;
2.  import java.io.Serializable;
3.
4.  public class Account implements Serializable {
5.      public ResultObj ResultObj;
6.      public static class ResultObj implements Serializable{
7.          public int UserID;
8.          public String UserName;
9.          public String Email;
10.         public String Telephone;
11.         public boolean Gender;
12.         public int CollegeID;
13.         public String CollegeName;
14.         public String RoleName;
15.         public int RoleID;
16.         public String AccessToken;
17.         public int AccessTokenErrCode;
18.         public String ReturnUrl;
19.         public String DataToken;
20.     }
21.     public int Status;
22.     public int StatusCode;
23.     public String Msg;
24.     public String ErrorObj;
25. }
```

第 16 行，存储返回的 AccessToken，这是我们最关心的数据。
第 21 行，存储响应的状态。
第 23 行，存储返回的具体消息内容。

6.4.3 登录逻辑

修改 LoginAbilitySlice.java，代码如下。

```
1.  package com.example.smartagriculture.slice;
2.
3.  import com.example.smartagriculture.ResourceTable;
4.  import com.example.smartagriculture.bean.Account;
5.  import com.example.smartagriculture.net.Wan;
6.  import ohos.aafwk.ability.AbilitySlice;
7.  import ohos.aafwk.content.Intent;
8.  import ohos.agp.components.Button;
9.  import ohos.agp.components.Component;
10. import ohos.agp.components.TextField;
11. import poerty.jianjian.converter.gson.GsonConverterFactory;
```

```
12.     import poetry.jianjia.Call;
13.     import poetry.jianjia.Callback;
14.     import poetry.jianjia.JianJia;
15.     import poetry.jianjia.Response;
16.
17.     public class LoginAbilitySlice extends AbilitySlice {
18.
19.         private Button btnLogin;
20.         private TextField tfAccount, tfPasswd;
21.         private JianJia mJianJia;
22.         private Wan mWan;
23.         private String TAG = "智慧农业";
24.         @Override
25.         protected void onStart(Intent intent) {
26.             super.onStart(intent);
27.             super.setUIContent(ResourceTable.Layout_ability_login);
28.             btnLogin = (Button) findComponentById(ResourceTable.Id_btnLogin);
29.             tfAccount = (TextField) findComponentById(ResourceTable.Id_tfAccount);
30.             tfPasswd = (TextField) findComponentById(ResourceTable.Id_tfPasswd);
31.
32.             btnLogin.setClickedListener(new Component.ClickedListener() {
33.                 @Override
34.                 public void onClick(Component component) {
35.                     login(tfAccount.getText().trim(), tfPasswd.getText().trim());
36.                 }
37.             });
38.             mJianJia = new JianJia.Builder()
39.                     .baseUrl("http://api.nlecloud.com")
40.                     .addConverterFactory(GsonConverterFactory.create())
41.                     .build();
42.             mWan = mJianJia.create(Wan.class);
43.         }
44.
45.         private void login(String account, String passwd) {
46.             System.out.println(TAG + "账号：" + account + "\n" + "密码：" + passwd);
47.             mWan.login(account, passwd, true).enqueue(new Callback<Account>() {
48.                 @Override
49.                 public void onResponse(Call<Account> call, Response<Account> response) {
50.                     try {
51.                         if(response.isSuccessful() && response.body().Status == 0){
52.                             System.out.println(TAG + "AccessToken: "+
```

```
response.body().ResultObj.AccessToken);
53.                        goToMainAbility();
54.                    } else {
55.                        System.out.println(TAG + "登录出错" + response.body().Msg);
56.                    }
57.
58.                } catch (Exception e) {
59.                    e.printStackTrace();
60.                }
61.
62.            }
63.
64.            @Override
65.            public void onFailure(Call<Account> call, Throwable throwable) {
66.                System.out.println(TAG + "登录请求出错"+ throwable.getMessage());
67.            }
68.        });
69.    }
70.
71.    private void goToMainAbility() {
72.        System.out.println(TAG + "前往主界面");
73.    }
74. }
```

第 21 行，创建 JianJia 类对象 mJianJia。

第 22 行，创建 Wan 类对象 mWan。

第 38~41 行，创建 JianJia 对象，传递新大陆物联网云平台的域名，设置 gson 转换器。

第 42 行，创建访问接口对象，以下使用该接口进行具体访问。

第 45 行，在第 34 行中单击事件发生后，程序执行到 login 处理流程，在该流程中，第 46 行，打印传递过来的账号和密码。

第 47~68 行，调用 mWan 的 login 方法，传递账号、密码和 true 进去，然后继续调用 enqueue 方法，在其中，创建了回调函数，在回调函数中，分别处理 onResponse（成功）响应和 onFailure（失败）响应。

第 49~62 行，在访问成功后，进入 onResponse 处理流程，判断具体的返回代码，如果为 0，代表服务器返回成功，此处可以解析里面的数据，可以使用 gson 转换器自动把返回结果转换成 Account 对象。兼葭支持添加数据转换器，在创建对象的时候添加数据转换器，也就是把 gson 添加进来。在 onResponse 方法里面就可以直接得到实体类对象了。为了实现这个功能，在 LoginAbilitySlice.java 代码第 40 行，在创建 JianJia 对象时添加了 gson 转换器，同时，还需要创建与登录响应内容相对应的 Account Bean。

第 52 行，使用 "response.body().ResultObj.AccessToken" 直接打印返回的 AccessToken。

第 53 行，在获取到 AccessToken 后，调用 goToMainAbility()，前往主界面处理流程。

第 71 行，实现跳往主界面的逻辑，此处打印一下日志，具体功能留待后续章节实现。

6.4.4 编译运行

编译运行程序，输入新大陆物联网云平台的账号和密码，打印的日志如图6-4所示。

图6-4 成功登录云平台

如果输入的账号或者密码不正确，则日志打印如图6-5所示。

图6-5 登录云平台失败

6.5 提交代码到仓库

提交代码到版本库，并打上标签，具体流程如下。
1）通过"git status"查看当前目录状态。
2）通过"git add."将当前目录下所有改动加入暂存区。
3）通过"git commit -m 创建登录功能"，提交暂存区内容到本地版本库。
4）通过"git status"查看当前工作目录，可以看到目录干净。
5）通过"git tag -a task6 -m 创建登录功能"，创建任务6标签。
6）通过"git log --pretty=oneline"查看版本库日志，可以看到提交的最新内容，以及task6标签，即任务6的版本。

任务 7　　创建底部标签导航栏

任务概述

本任务主要完成从 SplashAbility 下的 LoginAbilitySlice 切换到 MainAbility 下的 MainAbilitySlice，通过完成不同 Page Ability 的切换，实现从登录界面跳转到主界面的功能。此外，布局主界面导航，采用底部 TabList，实现多个页签栏平排的效果，如图 7-1 所示，再结合 PageSlider，实现标签导航栏设计，其中 PageSlider 留待下一个任务实现。

图 7-1　底部导航栏

知识目标

- 掌握不同 Page Ability 的切换。
- 掌握 Intent 的使用。
- 了解 StackLayout 布局。
- 掌握 TabList 的使用。
- 了解 Component 组件。
- 了解 ScrollView UI 组件。

能力目标

- 能使用 Intent 进行不同 Page Ability 的切换。
- 能使用 TabList 设计多标签。

7.1　不同 Page Ability 的切换

AbilitySlice 作为 Page 的内部单元，以 Action 的形式对外暴露，因此可以通过配置 Intent 的 Action 导航到目标 AbilitySlice。Page 间的导航可以使用 startAbility()或 startAbilityForResult() 方法，获得返回结果的回调为 onAbilityResult()。在 Ability 中调用 setResult()可以设置返回结果。

7.1.1 掌握 Intent 意图

Intent 是对象之间传递信息的载体。例如，当一个 Ability 需要启动另一个 Ability 或者一个 AbilitySlice 需要导航到另一个 AbilitySlice 时，可以通过 Intent 指定启动目标的同时携带相关数据。Intent 的构成元素包括 Operation 与 Parameters，见表 7-1。

表 7-1 Intent 的构成元素

属性	子属性	描述
Operation	Action	表示动作，通常使用系统预置 Action，应用也可以自定义 Action。例如 IntentConstants.ACTION_HOME 表示返回桌面动作
	Entity	表示类别，通常使用系统预置 Entity，应用也可以自定义 Entity。例如 Intent.ENTITY_HOME 表示在桌面显示图标
	Uri	表示 Uri 描述。如果在 Intent 中指定了 Uri，则 Intent 将匹配指定的 Uri 信息，包括 scheme、schemeSpecificPart、authority 和 path 信息
	Flags	表示处理 Intent 的方式。例如 Intent.FLAG_ABILITY_CONTINUATION 标记在本地的一个 Ability 是否可以迁移到远端设备继续运行
	BundleName	表示描述。如果在 Intent 中同时指定了 BundleName 和 AbilityName，则 Intent 可以直接匹配指定的 Ability
	AbilityName	表示待启动的 Ability 名称。如果在 Intent 中同时指定了 BundleName 和 AbilityName，则 Intent 可以直接匹配指定的 Ability
	DeviceId	表示运行指定 Ability 的设备 ID
Parameters	-	Parameters 是一种支持自定义的数据结构，开发者可以通过 Parameters 传递某些请求所需的额外信息

Intent 必须先使用 Operation 来设置属性。如果需要新增或修改属性，必须在设置 Operation 后再执行操作。

当 Intent 被用于发起请求时，根据指定元素的不同，分为两种类型。

1）如果同时指定了 BundleName 与 AbilityName，则根据 Ability 的全称（例如 "com.demoapp.FooAbility"）来直接启动应用。

2）如果未同时指定 BundleName 和 AbilityName，则根据 Operation 中的其他属性来启动应用。

7.1.2 根据 Ability 的全称启动应用

通过构造包含 BundleName 与 AbilityName 的 Operation 对象，可以启动一个 Ability，并导航到该 Ability，示例代码如下。

```
1.   Intent intent = new Intent();
2.
3.   // 通过 Intent 中的 OperationBuilder 类构造 operation 对象，指定设备标识（空串表示当前设备）、应用包名、Ability 名称
4.   Operation operation = new Intent.OperationBuilder()
5.           .withDeviceId("")
6.           .withBundleName("com.demoapp")
7.           .withAbilityName("com.demoapp.FooAbility")
8.           .build();
9.
10.  // 把 operation 设置到 intent 中
11.  intent.setOperation(operation);
```

```
12.     startAbility(intent);
```

第 4~8 行，创建 Operation 对象，并设置 BundleName 与 AbilityName。
第 11 行，将 Operation 对象绑定到 Intent 对象。
第 12 行，使用 startAbility 启动目标页面。

作为处理请求的对象，Ability 会在相应的回调方法中接收请求方传递的 Intent 对象。以导航到另一个 Ability 为例，导航的目标 Ability 可以在其 onStart()回调的参数中获得 Intent 对象。

7.1.3 根据 Operation 的其他属性启动应用

在某些场景下，开发者需要在应用中使用其他应用提供的某种能力，而不感知提供该能力的具体是哪一个应用。例如，开发者需要通过浏览器打开一个链接，而不关心用户最终选择哪一个浏览器应用，则可以通过 Operation 的其他属性（除 BundleName 与 AbilityName 之外的属性）描述需要的能力。如果设备上存在提供同一种能力的多个应用，则系统弹出候选列表，由用户选择由哪个应用处理请求。以下示例展示了如何使用 Intent 跨 Ability 查询天气信息。

1. 请求方

在 Ability 中构造 Intent 以及包含 Action 的 Operation 对象，并调用 startAbilityForResult()方法发起请求。然后重写 onAbilityResult()回调方法，对请求结果进行处理。

```
1.  private void queryWeather() {
2.      Intent intent = new Intent();
3.      Operation operation = new Intent.OperationBuilder()
4.              .withAction(Intent.ACTION_QUERY_WEATHER)
5.              .build();
6.      intent.setOperation(operation);
7.      startAbilityForResult(intent, REQ_CODE_QUERY_WEATHER);
8.  }
9.
10. @Override
11. protected void onAbilityResult(int requestCode, int resultCode, Intent resultData) {
12.     switch (requestCode) {
13.         case REQ_CODE_QUERY_WEATHER:
14.             // Do something with result.
15.             ...
16.             return;
17.         default:
18.             ...
19.     }
20. }
```

第 7 行，使用 startAbilityForResult 可以对返回结果进行 onAbilityResult 回调处理。
第 11~20 行，对返回结果进行处理。根据 requestCode 可以判断是来自哪个页面的返回结果。

2. 处理方

1）作为处理请求的对象，首先需要在配置文件中声明自身对外提供的能力，以便系统据此找到它并将其作为候选的请求处理者。

```
1.  {
2.      "module": {
3.          ...
4.          "abilities": [
5.              {
6.                  ...
7.                  "skills":[
8.                      {
9.                          "actions":[
10.                             "ability.intent.QUERY_WEATHER"
11.                         ]
12.                     }
13.                 ]
14.                 ...
15.             }
16.         ]
17.         ...
18.     }
19.     ...
20. }
```

第 10 行，对外声明可以提供的能力。

2）在 Ability 中配置路由以便支持以此 action 导航到对应的 AbilitySlice。

```
1.  @Override
2.  protected void onStart(Intent intent) {
3.      ...
4.      addActionRoute(Intent.ACTION_QUERY_WEATHER, DemoSlice.class.getName());
5.      ...
6.  }
```

第 4 行，在 Page Ability 中添加路由，以便支持特定的 action 导航到该 Ability 下的具体的 AbilitySlice。

3）在 Ability 中处理请求，并调用 setResult()方法暂存返回结果。

```
1.  @Override
2.  protected void onActive() {
3.      ...
4.      Intent resultIntent = new Intent();
5.      setResult(0, resultIntent);    //0 为当前 Ability 销毁后返回的 resultCode。
6.      ...
7.  }
```

第 5 行，设置即将返回的数据。其中第一个参数，即和请求方的 resultCode 对应。

7.1.4 LoginAbilitySlice 切换到 MainAbilitySlice

构造包含 BundleName 与 AbilityName 的 Operation 对象，启动主界面，此处 BundleName

为 com.example.smartagriculture，AbilityName 为 com.example.smartagriculture.MainAbility，修改 LoginAbilitySlice 文件的 goToMainAbility 方法，代码如下。

```
1.   private void goToMainAbility() {
2.       System.out.println(TAG + "前往主界面");
3.       Intent intent = new Intent();
4.       // 指定待启动FA的bundleName和abilityName
5.       Operation operation = new Intent.OperationBuilder()
6.               .withDeviceId("")
7.               .withBundleName("com.example.smartagriculture")
8.               .withAbilityName("com.example.smartagriculture.MainAbility")
9.               .build();
10.      intent.setOperation(operation);
11.      startAbility(intent);
12.  }
```

第 3 行，创建 Intent 对象。

第 5~9 行，设置 Operation 对象的属性，此处绑定了 BundleName 和 AbilityName，即根据 Ability 的全称启动主界面。

第 10 行，将 Operation 对象设置到 Intent 对象里。

第 11 行，使用 startAbility 传递 Intent 对象作为参数，启动另一个 Page Ability 的页面。

7.1.5 编译运行

使用远程模拟器进行调试。启动远程模拟器（可能需要登录授权），然后编译运行程序，如图 7-2 所示，模拟器界面显示已经成功跳转到主界面，同时日志区域打印了相关成功登录的流程。

图 7-2 登录界面跳转到主界面

7.1.6 提交代码到仓库

将此功能的代码提交到本地仓库,并添加日志"SplashAbility 下面的 LoginAbilitySlice 登录后跳转到 MainAbility 下面的 MainAbilitySlice"。

7.2 使用 TabList 设置多标签

TabList 的公有 XML 属性继承自 ScrollView。而 ScrollView 继承自 StackLayout,StackLayout 又继承自 Component,整个继承关系如图 7-3 所示。

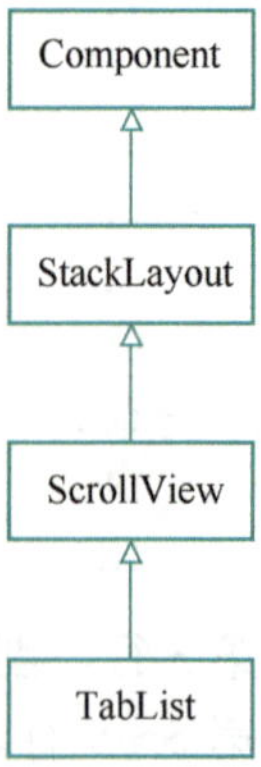

7.2.1 Component

ScrollView 是一种带滚动功能的组件,它采用滑动的方式在有限的区域内显示更多的内容。Component XML 常见属性如表 7-2 所示,其他属性可以参考官方指南。

图 7-3 TabList 继承图

表 7-2 Component XML 常见属性

属性分类	属性名称	中文描述	取值	取值说明	使用案例
基础属性	id	控件 identity,用以识别不同控件对象,每个控件唯一	integer 类型	仅可用于配置控件的 id	ohos:id="$+id:component_id"
	theme	样式	引用	仅可引用 pattern 资源	ohos:theme="$pattern:button_pattern"
	width	宽度,必填项	float 类型	可以配置表示尺寸的 float 类型。可以是浮点数值,其默认单位为 px;也可以是带 px、vp、fp 单位的浮点数值;也可以引用 float 资源	ohos:width="20" ohos:width="20vp" ohos:width="$float:size_value"
			match_parent	表示控件宽度与其父控件去掉内部边距后的宽度相同	ohos:width="match_parent"
			match_content	表示控件宽度由其包含的内容决定,包括其内容的宽度以及内部边距的总和	ohos:width="match_content"
	height	高度,必填项	float 类型	可以配置表示尺寸的 float 类型,同 width	ohos:height="20" ohos:height="20vp" ohos:height="$float:size_value"
			match_parent	表示控件高度与其父控件去掉内部边距后的高度相同	ohos:height="match_parent"
			match_content	表示控件高度由其包含的内容决定,包括其内容的高度以及内部边距的总和	ohos:height="match_content"

（续）

属性分类	属性名称	中文描述	取值	取值说明	使用案例
间距	padding	内间距	float 类型	表示间距尺寸的 float 类型。可以是浮点数值，其默认单位为 px；也可以是带 px、vp、fp 单位的浮点数值；也可以引用 float 资源。padding 与 left_padding、right_padding、start_padding、end_padding、top_padding、bottom_padding 属性有冲突，不建议一起使用。 在同时配置时，left_padding、right_padding、start_padding、end_padding、top_padding、bottom_padding 优先级高于 padding 属性	ohos:padding="20" ohos:padding="20vp" ohos:padding="$float:padding_value"
	left_padding	左内间距	float 类型	表示间距尺寸的 float 类型。取值与 padding 属性相同	ohos:left_padding="20" ohos:left_padding="20vp" ohos:left_padding="$float:padding_value"
	right_padding	右内间距	float 类型	表示间距尺寸的 float 类型。取值与 padding 属性相同	ohos:right_padding="20" ohos:right_padding="20vp" ohos:right_padding="$float:padding_value"
	top_padding	上内间距	float 类型	表示间距尺寸的 float 类型。取值与 padding 属性相同	ohos:top_padding="20" ohos:top_padding="20vp" ohos:top_padding="$float:padding_value"
	bottom_padding	下内间距	float 类型	表示间距尺寸的 float 类型。取值与 padding 属性相同	ohos:bottom_padding="20" ohos:bottom_padding="20vp" ohos:bottom_padding="$float:padding_value"
	margin	外边距	float 类型	表示间距尺寸的 float 类型。可以是浮点数值，其默认单位为 px；也可以是带 px、vp、fp 单位的浮点数值；也可以引用 float 资源。margin 与 left_margin、right_margin、start_margin、end_margin、top_margin、bottom_margin 属性有冲突，不建议一起使用。 同时配置时，margin 优先级高于 left_margin、right_margin、start_margin、end_margin、top_margin、bottom_margin 属性	ohos:margin="20" ohos:margin="20vp" ohos:margin="$float:margin_value"
	left_margin	左外边距	float 类型	表示间距尺寸的 float 类型。取值与 margin 属性相同	ohos:left_margin="20" ohos:left_margin="20vp" ohos:left_margin="$float:margin_value"
	right_margin	右外边距	float 类型	表示间距尺寸的 float 类型。取值与 margin 属性相同	ohos:right_margin="20" ohos:right_margin="20vp" ohos:right_margin="$float:margin_value"
	top_margin	上外边距	float 类型	表示间距尺寸的 float 类型。取值与 margin 属性相同	ohos:top_margin="20" ohos:top_margin="20vp" ohos:top_margin="$float:margin_value"
	bottom_margin	下外边距	float 类型	表示间距尺寸的 float 类型。取值与 margin 属性相同	ohos:bottom_margin="20" ohos:bottom_margin="20vp" ohos:bottom_margin="$float:margin_value"

7.2.2 StackLayout

StackLayout 直接在屏幕上开辟出一块空白的区域,添加到这个布局区域中的视图都是以层叠的方式显示,而它会把这些视图默认放到这块区域的左上角,第一个添加到布局区域中的视图显示在最底层,最后一个被添加的放在最顶层。上一层的视图会覆盖下一层的视图,如图 7-4 所示。

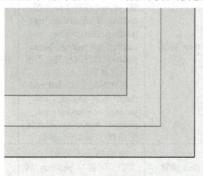

图 7-4　StackLayout 示意图

StackLayout 没有自有的 XML 属性,公有 XML 属性继承自 Component。StackLayout 所包含组件可支持的 XML 属性如表 7-3 所示。

表 7-3　StackLayout 所包含组件可支持的 XML 属性

属性名称	中文描述	取值	取值说明	使用案例
layout_alignment	对齐方式	left	表示左对齐	ohos:layout_alignment="top" ohos:layout_alignment="top|left" 可以设置取值项如表中所列,也可以使用"|"进行多项组合
		top	表示顶部对齐	
		right	表示右对齐	
		bottom	表示底部对齐	
		horizontal_center	表示水平居中对齐	
		vertical_center	表示垂直居中对齐	
		center	表示居中对齐	

7.2.3 ScrollView

ScrollView 是一种带滚动功能的组件,它采用滑动的方式在有限的区域内显示更多的内容。ScrollView 的公有 XML 属性继承自 StackLayout。ScrollView 的自有 XML 属性如表 7-4 所示。

表 7-4　ScrollView 的自有 XML 属性

属性名称	中文描述	取值	取值说明	使用案例
match_viewport	是否拉伸匹配	boolean 类型	可以直接设置 true 或 false,也可以引用 boolean 资源	ohos:match_viewport="true" ohos:match_viewport="$boolean:true"
rebound_effect	回弹效果	boolean 类型	可以直接设置 true 或 false,也可以引用 boolean 资源	ohos:rebound_effect="true" ohos:rebound_effect="$boolean:true"

7.2.4 TabList

TabList 可以实现多个页签栏间的切换,Tab 为某个页签。子页签通常放在内容区上方,展示不同的分类。页签名称应该简洁明了，清晰描述分类的内容。TabList 的公有 XML 属性继承自 ScrollView。TabList 的自有 XML 属性如表 7-5 所示。

表 7-5 TabList 的自有 XML 属性

属性名称	中文描述	取值	取值说明	使用案例
fixed_mode	固定所有页签并同时显示	boolean 类型	可以直接设置 true 或 false，也可以引用 boolean 资源	ohos:fixed_mode="true" ohos:fixed_mode="$boolean:true_tag"
orientation	页签排列方向	horizontal	表示水平排列	ohos:orientation="horizontal"
		vertical	表示垂直排列	ohos:orientation="vertical"
normal_text_color	未选中的文本颜色	color 类型	可以直接设置色值，也可以引用 color 资源	ohos:normal_text_color="#FFFFFFFF" ohos:normal_text_color="$color:black"
selected_text_color	选中的文本颜色	color 类型	可以直接设置色值，也可以引用 color 资源	ohos:selected_text_color="#FFFFFFFF" ohos:selected_text_color="$color:black"
selected_tab_indicator_color	选中页签的颜色	color 类型	可以直接设置色值，也可以引用 color 资源	ohos:selected_tab_indicator_color="#FFFFFFFF" ohos:selected_tab_indicator_color="$color:black"
selected_tab_indicator_height	选中页签的高度	float 类型	表示尺寸的 float 类型。可以是浮点数值，其默认单位为 px；也可以是带 px、vp、fp 单位的浮点数值；也可以引用 float 资源	ohos:selected_tab_indicator_height="100" ohos:selected_tab_indicator_height="20vp" ohos:selected_tab_indicator_height="$float:size_value"
tab_indicator_type	页签指示类型	invisible	表示选中的页签无指示标记	ohos:tab_indicator_type="invisible"
		bottom_line	表示选中的页签通过底部下画线标记	ohos:tab_indicator_type="bottom_line"
		left_line	表示选中的页签通过左侧分割线标记	ohos:tab_indicator_type="left_line"
		oval	表示选中的页签通过椭圆背景标记	ohos:tab_indicator_type="oval"
tab_length	页签长度	float 类型	表示尺寸的 float 类型。可以是浮点数值，其默认单位为 px；也可以是带 px、vp、fp 单位的浮点数值；也可以引用 float 资源	ohos:tab_length="100" ohos:tab_length="20vp" ohos:tab_length="$float:size_value"
tab_margin	页签间距	float 类型	表示尺寸的 float 类型。可以是浮点数值，其默认单位为 px；也可以是带 px、vp、fp 单位的浮点数值；也可以引用 float 资源	ohos:tab_margin="100" ohos:tab_margin="20vp" ohos:tab_margin="$float:size_value"
text_alignment	文本对齐方式	left	表示文本靠左对齐。可以设置取值项如表中所列，也可以使用"|"进行多项组合	ohos:text_alignment="center" ohos:text_alignment="top|left"
		top	表示文本靠顶部对齐	
		right	表示文本靠右对齐	
		bottom	表示文本靠底部对齐	
		horizontal_center	表示文本水平居中对齐	
		vertical_center	表示文本垂直居中对齐	
		center	表示文本居中对齐	
		start	表示文本靠起始端对齐	
		end	表示文本靠结尾端对齐	
text_size	文本大小	float 类型	表示尺寸的 float 类型。可以是浮点数值，其默认单位为 px；也可以是带 px、vp、fp 单位的浮点数值；也可以引用 float 资源	ohos:text_size="100" ohos:text_size="16fp" ohos:text_size="$float:size_value"

7.2.5 实现 TabList 功能

为了实现 TabList 功能，需要修改三个文件：ability_main.xml、MainAbilitySlice.java 和 string.json。

1. 更新 ability_main.xml

```
1.   <?xml version="1.0" encoding="utf-8"?>
2.   <DependentLayout
3.       xmlns:ohos="http://schemas.huawei.com/res/ohos"
4.       ohos:height="match_parent"
5.       ohos:width="match_parent"
6.       ohos:background_element="$graphic:background_ability_main">
7.
8.       <TabList
9.           ohos:id="$+id:tltNav"
10.          ohos:top_margin="40vp"
11.          ohos:tab_margin="24vp"
12.          ohos:tab_length="140vp"
13.          ohos:text_size="20fp"
14.          ohos:height="50vp"
15.          ohos:width="match_parent"
16.          ohos:layout_alignment="center"
17.          ohos:orientation="horizontal"
18.          ohos:text_alignment="center"
19.          ohos:normal_text_color="#999999"
20.          ohos:selected_text_color="#FF156E52"
21.          ohos:selected_tab_indicator_color="#FF156E52"
22.          ohos:selected_tab_indicator_height="2vp"
23.          ohos:align_parent_bottom="true"
24.          ohos:fixed_mode="true"
25.          ohos:rebound_effect="true"/>
26.
27.  </DependentLayout>
```

第 8 行，增加 TabList 组件。

第 9 行，设置 TabList 的 id 为 tltNav。

第 10 行，设置 TabList 的上外边距为 40vp。

第 11 行，设置 TabList 的页签间距为 24vp。

第 12 行，设置 TabList 的页签长度为 140vp。

第 13 行，设置 TabList 的页签的文本大小为 20fp。

第 14 行，设置 TabList 的高度为 50vp。

第 15 行，设置 TabList 的宽度为父布局宽度。

第 16 行，设置 TabList 居中。

第 17 行，设置 TabList 的页签排列方向为水平方向。

第 18 行，设置文本对齐方式居中。

第 19 行，设置文本未选中时的颜色为#999999。

第 20 行，设置文本选中时的颜色为#FF156E52。
第 21 行，设置选中页签的颜色为#FF156E52。
第 22 行，设置选中页签的高度为 2vp。
第 23 行，设置 TabList 位于布局底部。
第 24 行，设置固定所有页签并同时显示。
第 25 行，设置回弹效果，该属性继承自 ScrollView。

2. 更新 MainAbilitySlice.java

```
1.   package com.example.smartagriculture.slice;
2.
3.   import com.example.smartagriculture.ResourceTable;
4.   import ohos.aafwk.ability.AbilitySlice;
5.   import ohos.aafwk.content.Intent;
6.   import ohos.agp.components.TabList;
7.
8.   import java.util.ArrayList;
9.   import java.util.List;
10.
11.  public class MainAbilitySlice extends AbilitySlice {
12.      private List<TabList.Tab> tabs;
13.      private String TAG = "智慧农业";
14.      @Override
15.      public void onStart(Intent intent) {
16.          super.onStart(intent);
17.          super.setUIContent(ResourceTable.Layout_ability_main);
18.          initTabList();
19.      }
20.
21.      private void initTabList() {
22.          tabs = new ArrayList<TabList.Tab>();
23.          TabList TabList = (TabList) findComponentById(ResourceTable.Id_tltNav);
24.          TabList.Tab tabAir = TabList.new Tab(getContext());
25.          tabAir.setText(ResourceTable.String_strPgAir);
26.          TabList.addTab(tabAir);
27.          tabs.add(tabAir);
28.
29.          TabList.Tab tabSoil = TabList.new Tab(getContext());
30.          tabSoil.setText(ResourceTable.String_strPgSoil);
31.          TabList.addTab(tabSoil);
32.          tabs.add(tabSoil);
33.
34.          TabList.Tab tabControl = TabList.new Tab(getContext());
35.          tabControl.setText(ResourceTable.String_strPgControl);
36.          TabList.addTab(tabControl);
37.          tabs.add(tabControl);
```

```
38.
39.            TabList.Tab tabMe = TabList.new Tab(getContext());
40.            tabMe.setText(ResourceTable.String_strPgMe);
41.            TabList.addTab(tabMe);
42.            tabs.add(tabMe);
43.
44.            tabAir.select();
45.            TabList.addTabSelectedListener(new TabList.TabSelectedListener() {
46.                @Override
47.                public void onSelected(TabList.Tab tab) {
48.                    int i = tab.getPosition();
49.                    System.out.println(TAG + "这是第" + i + "个 Tab");
50.                    System.out.println(TAG + "Tab: " + tab.getText());
51.                }
52.
53.                @Override
54.                public void onUnselected(TabList.Tab tab) {
55.
56.                }
57.
58.                @Override
59.                public void onReselected(TabList.Tab tab) {
60.
61.                }
62.            });
63.        }
64.
65.        @Override
66.        public void onActive() {
67.            super.onActive();
68.        }
69.
70.        @Override
71.        public void onForeground(Intent intent) {
72.            super.onForeground(intent);
73.        }
74. }
```

第 12 行，声明 TabList.Tab 的集合 tabs。

第 18 行，调用 initTabList 方法，初始化 Tab 页签。

第 21～63 行，initTabList 方法的实现。

第 22 行，创建 TabList.Tab 集合，赋值给 tabs。

第 23 行，获取 ability_main.xml 的 TabList 的组件 tltNav，赋值给 TabList。

第 24 行，创建 TabList.Tab 对象 tabAir。

第 25 行，设置 tabAir 的页签文本，该文本引用自 base/element/string.json。

第 26 行，将 tabAir 添加到 TabList 集合里，即添加到页面 TabList 里。

第 29～42 行，利用与第 26 行同样的方法，创建 tabSoil、tabControl、tabMe 页签。

第 44 行，设置默认选中的页签为 tabAir。

第 45 行，设置 TabList 响应焦点变化。

第 47～51 行，设置选中页签的响应事件处理，此处首先获取了单击的页签的位置，从 0 计数，然后打印页签的位置和页签的文本内容。

3．增加 base/element/string.json 文件字符串

```
1.  {
2.    "name": "strPgSoil",
3.    "value": "土壤"
4.  },
5.  {
6.    "name": "strPgAir",
7.    "value": "大气"
8.  },
9.  {
10.   "name": "strPgControl",
11.   "value": "控制"
12.  },
13.  {
14.   "name": "strPgMe",
15.   "value": "我的"
16.  }
```

此次增加了 **strPgSoil**、**strPgAir**、**strPgControl**、**strPgMe** 字符串。

7.2.6 编译运行

使用远程模拟器调试程序，编译运行，输入新大陆物联网云平台账号和密码，运行结果如图 7-5 所示，单击页签，可以在日志区看到与页签相关的具体位置和打印的文本内容。

图 7-5　TabList 功能实现

7.3 提交代码到仓库

将任务 7 的功能提交到仓库，具体流程如下。

1）通过"git status"查看当前目录状态。

2）通过"git add."将当前目录下所有改动加入暂存区。

3）通过"git commit -m 创建底部标签导航栏"，提交暂存区内容到本地版本库。

4）通过"git status"查看当前工作目录，可以看到目录干净。

5）通过"git tag -a task7 -m 创建底部标签导航栏"，创建任务 7 标签。

6）通过"git log --pretty=oneline"查看版本库日志，可以看到提交的最新内容，以及"task7"标签，即任务 7 的版本。

任务 8　创建大气环境监控界面

任务概述

本任务主要完成大气监控界面设计,界面功能主要向用户显示各类传感器的数据,如图 8-1 所示,包括了温度、湿度、风速、风向、光照、气压、PM2.5 和二氧化碳。同时,在标签导航栏上部显示最新数据同步的时间。

图 8-1　大气环境监控

知识目标

- 掌握 PageSlider 组件。
- 掌握布局和组件的常见属性。
- 掌握 DependentLayout 和 DirectionalLayout 的嵌套设计。

能力目标

- 能使用 PageSilder 组件进行多页面设计。
- 能灵活使用 DependentLayout 和 DirectionalLayout 嵌套设计。

8.1 使用 PageSlider 组件切换页面

PageSlider 是用于切换页面的组件,它通过响应滑动事件完成页面间的切换。PageSlider 没有自有的 XML 属性,公有 XML 属性继承自 StackLayout。

8.1.1 增加 PageSlider

在 ability_main.xml 中增加 PageSlider 组件，修改代码如下。

```
1.  <PageSlider
2.      ohos:id="$+id:page_sliderMain"
3.      ohos:height="match_parent"
4.      ohos:width="match_parent"
5.      ohos:layout_alignment="horizontal_center"
6.      ohos:above="$id:tltNav"/>
```

第 2 行，设置 PageSlider 的 id。
第 3~4 行，设置高宽匹配父布局。
第 5 行，设置 PageSlider 水平居中。
第 6 行，设置 PageSilder 组件位于 TabList 组件的上方。

8.1.2 创建 PageSliderProvider 子类

每个页面可能需要呈现不同的数据，因此要想适配不同的数据结构，就必须创建类，继承 PageSliderProvider.java，重写如表 8-1 所示方法。

表 8-1 PageSliderProvider 方法

方法名	作用
getCount()	获取可用视图的数量
createPageInContainer(ComponentContainer componentContainer, int position)	在指定位置创建页面
destroyPageFromContainer(ComponentContainer componentContainer, int i, Object o)	销毁容器中的指定页面
isPageMatchToObject(Component component, Object o)	视图是否关联指定对象

在 com.example.smartagriculture.provider 包下创建 PagesProvider.java，并继承 PageSliderProvider.java。PagesProvider.java 的完整代码如下。

```java
1.  package com.example.smartagriculture.provider;
2.
3.  import ohos.agp.components.Component;
4.  import ohos.agp.components.ComponentContainer;
5.  import ohos.agp.components.PageSliderProvider;
6.
7.  import java.util.List;
8.
9.  public class PagesProvider extends PageSliderProvider {
10.     private List<Component> pages;
11.
12.     public PagesProvider(List<Component> pages) {
13.         this.pages = pages;
14.     }
15.
```

```
16.     @Override
17.     public int getCount() {
18.         return pages.size();
19.     }
20.
21.     @Override
22.     public Object createPageInContainer(ComponentContainer component-
Container, int i) {
23.         componentContainer.addComponent(pages.get(i));
24.         return pages.get(i);
25.     }
26.
27.     @Override
28.     public void destroyPageFromContainer(ComponentContainer componentContainer,
int i, Object o) {
29.         componentContainer.removeComponent(pages.get(i));
30.     }
31.
32.     @Override
33.     public boolean isPageMatchToObject(Component component, Object o) {
34.         return component == o;
35.     }
36. }
```

第 10 行，声明数据源，此处数据源是页面组件的集合。

第 12 行，构造方法，传递数据源参数，初始化数据源。

第 17 行，返回可用的页面数目。

第 22～25 行，在指定位置创建页面。

第 28～30 行，销毁容器中指定的页面。

第 33～35 行，判断视图是否关联指定对象。

8.2 大气监控界面设计

1. 布局设计

在 base/layout 布局目录下，创建布局文件 pagesilder_air.xml 用于大气监控界面的布局设计。

大气环境监控界面的嵌套布局如图 8-2 所示，最外层是依赖布局，传感器数据行与行之间是垂直的方向布局，传感器数据行是水平的方向布局。完整代码可以参考本书配套代码，在代码目录通过以下命令提取任务 8 的代码版本。

```
git checkout task8
```

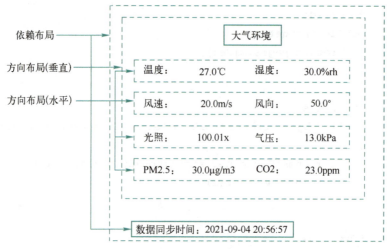

图 8-2　嵌套布局

2. 引用字符串

更新 string.json 文件，路径为 base/element/string.json，增加内容，如表 8-2 所示。

表 8-2　字符串表

name	value	name	value	name	value
strTxtAirEnvironment	大气环境	strTxtTemperature	温度：	strTxtHumidity	湿度：
strTxtWindSpeed	风速：	strTxtWindDirection	风向：	strTxtIllumination	光照：
strTxtAtmos	气压：	strTxtPM2_5	PM2.5：	strTxtCarbonDioxide	CO2：
strTxtDataSyncDatetime	数据同步时间：				

8.3　更新 MainAbilitySlice.java

在 MainAbilitySlice 中，增加 PageSlider 页面，并将大气环境监控界面加入 PageSlider 组件。

8.3.1　PageSlider 常用方法

PageSlider 常用方法如表 8-3 所示。

表 8-3　PageSlider 常用方法

方法名	作用
setProvider(PageSliderProvider provider)	设置 Provider，用于配置 PageSlider 的数据结构
addPageChangedListener(PageChangedListener listener)	响应页面切换事件
removePageChangedListener(PageChangedListener listener)	移除页面切换的响应
setOrientation(int orientation)	设置布局方向
setPageCacheSize(int count)	设置要保留当前页面两侧的页面数
setCurrentPage(int itemPos)	设置当前展示页面
setCurrentPage(int itemPos, boolean smoothScroll)	设置当前展示界面，并确定是否需要平滑滚动

(续)

方法名	作用
setSlidingPossible(boolean enable)	是否启用页面滑动
setReboundEffect(boolean enabled)	是否启用回弹效果
setReboundEffectParams(int overscrollPercent, float overscrollRate,int remainVisiblePercent)	配置回弹效果参数
setReboundEffectParams(ReboundEffectParams reboundEffectParams)	
setPageSwitchTime(int durationMs)	设置页面切换时间

8.3.2 更新 MainAbilitySlice.java 代码

将 pagesilder_air.xml 大气监控界面设置到 PageSlider 组件里，同时绑定 TabList 页签，使得单击大气页签时，可以使用 PageSlider 组件创建大气监控界面。MainAbilitySlice.java 完整代码如下。

```
1.  package com.example.smartagriculture.slice;
2.
3.  import com.example.smartagriculture.ResourceTable;
4.  import com.example.smartagriculture.provider.PagesProvider;
5.  import ohos.aafwk.ability.AbilitySlice;
6.  import ohos.aafwk.content.Intent;
7.  import ohos.agp.components.*;
8.
9.  import java.util.ArrayList;
10. import java.util.List;
11.
12. public class MainAbilitySlice extends AbilitySlice {
13.     private List<TabList.Tab> tabs;
14.     private String TAG = "智慧农业";
15.     private PageSlider pageSlider;
16.     private List<Component> pages;
17.     private DependentLayout dependentLayoutPagesliderAir;
18.     @Override
19.     public void onStart(Intent intent) {
20.         super.onStart(intent);
21.         super.setUIContent(ResourceTable.Layout_ability_main);
22.         initTabList();
23.         initPageSlider();
24.     }
25.
26.     private void initTabList() {
27.         tabs = new ArrayList<TabList.Tab>();
28.         TabList TabList = (TabList) findComponentById(ResourceTable.Id_tltNav);
29.         TabList.Tab tabAir = TabList.new Tab(getContext());
30.         tabAir.setText(ResourceTable.String_strPgAir);
31.         TabList.addTab(tabAir);
32.         tabs.add(tabAir);
33.
```

```
34.        TabList.Tab tabSoil = TabList.new Tab(getContext());
35.        tabSoil.setText(ResourceTable.String_strPgSoil);
36.        TabList.addTab(tabSoil);
37.        tabs.add(tabSoil);
38.
39.        TabList.Tab tabControl = TabList.new Tab(getContext());
40.        tabControl.setText(ResourceTable.String_strPgControl);
41.        TabList.addTab(tabControl);
42.        tabs.add(tabControl);
43.
44.        TabList.Tab tabMe = TabList.new Tab(getContext());
45.        tabMe.setText(ResourceTable.String_strPgMe);
46.        TabList.addTab(tabMe);
47.        tabs.add(tabMe);
48.
49.        tabAir.select();
50.        TabList.addTabSelectedListener(new TabList.TabSelectedListener() {
51.            @Override
52.            public void onSelected(TabList.Tab tab) {
53.                int i = tab.getPosition();
54.                System.out.println(TAG + "这是第" + i + "个 Tab");
55.                System.out.println(TAG + "Tab: " + tab.getText());
56.                //平滑切换到 tab 对应的页面。
57.                pageSlider.setCurrentPage(i, true);
58.            }
59.
60.            @Override
61.            public void onUnselected(TabList.Tab tab) {
62.
63.            }
64.
65.            @Override
66.            public void onReselected(TabList.Tab tab) {
67.
68.            }
69.        });
70.    }
71.
72.    private void initPageSlider() {
73.        pageSlider = (PageSlider) findComponentById(ResourceTable.Id_page_sliderMain);
74.        LayoutScatter dc = LayoutScatter.getInstance(getContext());
75.        dependentLayoutPagesliderAir = (DependentLayout) dc.parse(ResourceTable.Layout_pageslider_air, null, false);
76.        pages = new ArrayList<Component>();
77.        pages.add(dependentLayoutPagesliderAir);
78.        pageSlider.setProvider(new PagesProvider(pages));
```

```
79.            pageSlider.addPageChangedListener(new PageSlider.
PageChangedListener() {
80.                @Override
81.                public void onPageSliding(int i, float v, int i1) {
82.
83.                }
84.
85.                @Override
86.                public void onPageSlideStateChanged(int i) {
87.
88.                }
89.
90.                @Override
91.                public void onPageChosen(int i) {
92.                    tabs.get(i).select();
93.                    System.out.println(TAG + "选中page " + i);
94.
95.                }
96.            });
97.        }
98.
99.        @Override
100.       public void onActive() {
101.           super.onActive();
102.       }
103.
104.       @Override
105.       public void onForeground(Intent intent) {
106.           super.onForeground(intent);
107.       }
108.    }
```

第 15 行，增加 PageSlider 成员。

第 16 行，增加页面组件列表。

第 17 行，增加依赖布局成员属性。

第 23 行，在 onStart 中调用 initPageSlider 方法。

第 57 行，当用户单击页签时，根据页签的位置 i，选择相对应的 PageSlider 里的页面，即可以将 TabList 与 PageSlider 绑定起来。

第 72~97 行，initPageSlider 方法的具体实现。

第 73 行，获取布局界面里的 PageSilder 组件并赋值给 pageSlider。

第 74、75 行，获取大气监控界面的布局并赋值给 dependentLayoutPagesliderAir。

第 76、77 行，创建 pages 列表，并将 dependentLayoutPagesliderAir 放入 pages 列表。

第 78 行，通过适配器构建大气监控界面并放入 PageSlider 里。

第 79 行，调用 pageSlider 的 addPageChangedListener 方法，创建监听器。

第 90~95 行，重写 onPageChosen 方法，与 TabList 绑定，当 PageSlider 页面滑动到其他页面时，返回此时页面的位置 i，tabs 根据位置 i 选择相对应的页签。

8.3.3 编译运行

连接远程模拟器，编译运行，结果如图 8-3 所示，大气页签显示大气环境监控界面。

图 8-3　大气监控界面

8.4 提交代码到仓库

将任务 8 的代码提交到本地仓库，完整流程如下。
1）通过"git status"查看当前目录状态。
2）通过"git add."将当前目录下所有改动加入暂存区。
3）通过"git commit -m 创建大气环境监控界面"，提交暂存区内容到本地版本库。
4）通过"git status"查看当前工作目录，可以看到目录干净。
5）通过"git tag -a task8 -m 创建大气环境监控界面"，创建任务 8 标签。
6）通过"git log --pretty=oneline"查看版本库日志，可以看到提交的最新内容，以及"task8"标签，即任务 8 的版本。

任务 9　　创建参数设置界面

任务概述

智慧农业应用不是一个封闭的软件系统，需要从外界获取数据，或者发送命令到外部系统，这样应用就少不了和外界的系统进行通信，所以一些参数必须事先设置好，才能保证通信的实现。例如，要从物联网云平台获取传感器数据和通过物联网云平台控制水阀的打开和关闭，App 就要和云平台进行通信，需要设置的参数有物联网云平台的 IP 地址、项目标识、传感器 ID 以及执行器的 ID。为了实现参数设置，任务 9 需要设计个人设置界面和云平台参数设置界面及实现组件基本的单击事件处理，界面如图 9-1 和图 9-2 所示。

图 9-1　个人设置　　　　　　　　　图 9-2　云平台参数设置

知识目标

- 掌握组件的单击事件处理。
- 掌握 Button 单击事件处理。

能力目标

- 能处理组件的单击事件。
- 能使用 Button 处理单击事件。

9.1　个人设置界面

在创建"我的"界面时，需要先创建"土壤"界面和"控制"界面，将它们放入 PageSlider，以达到页面的位置序号与 TabList 的序号相对应。如图 9-3 所示，在 TabList 里，"大气"位置为 0，"土壤"位置为 1，"控制"位置为 2，"我的"位置为 3，在 PageSlider 中按序添加页面，则可以将页面序号与页签序号一一对应。

大气　　土壤　　控制　　我的

图 9-3　导航栏

9.1.1　创建土壤界面

在 layout 目录下创建 pagesilder_soil.xml，完整代码如下。

```
1.  <?xml version="1.0" encoding="utf-8"?>
2.  <DependentLayout
3.      xmlns:ohos="http://schemas.huawei.com/res/ohos"
4.      ohos:height="match_parent"
5.      ohos:width="match_parent"
6.      ohos:orientation="vertical">
7.  
8.  </DependentLayout>
```

9.1.2　创建控制界面

在 layout 目录下创建 pagesilder_control.xml，完整代码如下。

```
1.  <?xml version="1.0" encoding="utf-8"?>
2.  <DirectionalLayout
3.      xmlns:ohos="http://schemas.huawei.com/res/ohos"
4.      ohos:height="match_parent"
5.      ohos:width="match_parent"
6.      ohos:orientation="vertical">
7.  
8.  </DirectionalLayout>
```

9.1.3　创建我的界面

1. 布局设计

在 layout 目录下创建 pagesilder_me.xml，代码如下。

```
1.  <?xml version="1.0" encoding="utf-8"?>
2.  <DirectionalLayout
3.      xmlns:ohos="http://schemas.huawei.com/res/ohos"
4.      ohos:height="match_parent"
5.      ohos:width="match_parent"
6.      ohos:orientation="vertical"
7.      ohos:background_element="$graphic:background_ability_main">
8.      <Text
9.          ohos:height="match_content"
10.         ohos:width="match_parent"
11.         ohos:top_margin="20vp"
12.         ohos:bottom_margin="20vp"
```

```
13.            ohos:text="$string:strTxtMe"
14.            ohos:text_size="30vp"
15.            ohos:layout_alignment="horizontal_center"
16.            ohos:text_alignment="center"/>
17.
18.     <DirectionalLayout
19.         ohos:height="match_content"
20.         ohos:width="match_parent"
21.         ohos:top_margin="15vp"
22.         ohos:bottom_margin="15vp"
23.         ohos:left_margin="15vp"
24.         ohos:right_margin="15vp"
25.         ohos:padding="5vp"
26.         ohos:background_element="$graphic:page_me_account_bg"
27.         ohos:orientation="vertical">
28.         <Text
29.             ohos:height="match_content"
30.             ohos:width="match_content"
31.             ohos:text="$string:strTxtAccount"
32.             ohos:text_size="20fp"
33.             ohos:text_color="#666"/>
34.         <DirectionalLayout
35.             ohos:height="match_content"
36.             ohos:width="match_content"
37.             ohos:orientation="horizontal"
38.             ohos:left_margin="15vp"
39.             ohos:top_margin="15vp">
40.             <Text
41.                 ohos:height="match_content"
42.                 ohos:width="match_content"
43.                 ohos:text="$string:strTxtTelephone"
44.                 ohos:text_size="18fp"
45.                 ohos:text_color="#666"/>
46.             <Text
47.                 ohos:id="$+id:txtTelephoneValue"
48.                 ohos:height="match_content"
49.                 ohos:width="match_content"
50.                 ohos:text="12345667890"
51.                 ohos:left_margin="3vp"
52.                 ohos:text_size="18fp"/>
53.         </DirectionalLayout>
54.         <DirectionalLayout
55.             ohos:height="match_content"
56.             ohos:width="match_content"
57.             ohos:orientation="horizontal"
58.             ohos:left_margin="15vp"
59.             ohos:top_margin="15vp">
60.             <Text
```

```
61.                    ohos:height="match_content"
62.                    ohos:width="match_content"
63.                    ohos:text="$string:strTxtAccessToken"
64.                    ohos:text_size="18fp"
65.                    ohos:text_color="#666"/>
66.                <Text
67.                    ohos:id="$+id:txtAccessTokenValue"
68.                    ohos:height="match_content"
69.                    ohos:width="match_content"
70.                    ohos:text="*****************************"
71.                    ohos:left_margin="3vp"
72.                    ohos:text_size="18fp"/>
73.            </DirectionalLayout>
74.        </DirectionalLayout>
75.
76.        <DirectionalLayout
77.            ohos:height="match_content"
78.            ohos:width="match_parent"
79.            ohos:top_margin="30vp"
80.            ohos:bottom_margin="15vp"
81.            ohos:left_margin="15vp"
82.            ohos:right_margin="15vp"
83.            ohos:padding="5vp"
84.            ohos:background_element="$graphic:page_me_setting_bg"
85.            ohos:orientation="vertical">
86.            <Text
87.                ohos:height="match_content"
88.                ohos:width="match_content"
89.                ohos:text="$string:strTxtSetting"
90.                ohos:text_size="20fp"
91.                ohos:text_color="#666"/>
92.            <Text
93.                ohos:id="$+id:txtCloudParameterSetting"
94.                ohos:height="match_content"
95.                ohos:width="match_content"
96.                ohos:text="$string:strTxtCloudParameterSetting"
97.                ohos:text_size="18fp"
98.                ohos:text_color="#666"
99.                ohos:left_margin="15vp"
100.                ohos:top_margin="15vp"/>
101.            <Text
102.                ohos:id="$+id:txtQuit"
103.                ohos:height="match_content"
104.                ohos:width="match_content"
105.                ohos:text="$string:strTxtQuit"
106.                ohos:text_size="18fp"
107.                ohos:text_color="#666"
108.                ohos:left_margin="15vp"
```

```
109.                    ohos:top_margin="15vp"/>
110.         </DirectionalLayout>
111. </DirectionalLayout>
```

充分利用嵌套布局设计，最外层是垂直方向的布局：个人设置、账号信息、设置。内层账号信息与设置也是垂直方向的布局，手机号和 AccessToken 是水平方向的布局。

2. 背景设计

1）在 graphic 目录下创建 page_me_account_bg.xml，代码如下。

```
1. <?xml version="1.0" encoding="UTF-8" ?>
2. <shape xmlns:ohos="http://schemas.huawei.com/res/ohos"
3.        ohos:shape="rectangle">
4.     <corners ohos:radius="20"/>
5.     <solid
6.         ohos:color="#FFDBE4FC"/>
7. </shape>
```

2）在 graphic 目录下创建 page_me_setting_bg.xml，代码如下。

```
1. <?xml version="1.0" encoding="UTF-8" ?>
2. <shape xmlns:ohos="http://schemas.huawei.com/res/ohos"
3.        ohos:shape="rectangle">
4.     <solid
5.         ohos:color="#FFDBE4FC"/>
6. </shape>
```

3. 引用字符串

在字符串引用文件 base/element/string.json 中增加新的字符串，代码如下。

```
1.  {
2.      "name": "strTxtAccount",
3.      "value": "账号信息"
4.  },
5.  {
6.      "name": "strTxtTelephone",
7.      "value": "手机号:"
8.  },
9.  {
10.     "name": "strTxtAccessToken",
11.     "value": "AccessToken:"
12. },
13. {
14.     "name": "strTxtSetting",
15.     "value": "设置"
16. },
17. {
18.     "name": "strTxtCloudParameterSetting",
19.     "value": "云平台参数设置"
20. },
```

```
21.    {
22.        "name": "strTxtQuit",
23.        "value": "退出当前账号"
24.    },
25.    {
26.        "name": "strTxtMe",
27.        "value": "个人设置"
28.    }
```

9.1.4 编辑 MainAbilitySlice.java

1. 增加成员属性

```
private DependentLayout dependentLayoutPagesliderSoil;
DirectionalLayout directionalLayoutPagesliderControl, directionalLayoutPagesliderMe;
```

2. 修改 initPageSlider 方法

```
1.  private void initPageSlider() {
2.      pageSlider = (PageSlider)findComponentById(ResourceTable.Id_page_sliderMain);
3.      LayoutScatter dc = LayoutScatter.getInstance(getContext());
4.      dependentLayoutPagesliderAir = (DependentLayout) dc.parse(ResourceTable.Layout_pageslider_air, null, false);
5.      dependentLayoutPagesliderSoil = (DependentLayout) dc.parse(ResourceTable.Layout_pageslider_soil, null, false);
6.      directionalLayoutPagesliderControl = (DirectionalLayout) dc.parse(ResourceTable.Layout_pageslider_control, null, false);
7.      directionalLayoutPagesliderMe = (DirectionalLayout) dc.parse(ResourceTable.Layout_pageslider_me, null, false);
8.      pages = new ArrayList<Component>();
9.      pages.add(dependentLayoutPagesliderAir);
10.     pages.add(dependentLayoutPagesliderSoil);
11.     pages.add(directionalLayoutPagesliderControl);
12.     pages.add(directionalLayoutPagesliderMe);
13.     pageSlider.setProvider(new PagesProvider(pages));
14.     pageSlider.addPageChangedListener(new PageSlider.PageChangedListener() {
15.         @Override
16.         public void onPageSliding(int i, float v, int i1) {
17.         }
18.         @Override
19.         public void onPageSlideStateChanged(int i) {
20.         }
21.         @Override
22.         public void onPageChosen(int i) {
23.             tabs.get(i).select();
24.             System.out.println(TAG + "选中page " + i);
```

```
25.            }
26.        });
27. }
```

第 5～7 行，增加 dependentLayoutPagesliderSoil、directional-LayoutPagesliderControl 和 directionalLayoutPagesliderMe 初始化 。

第 10～12 行，将三个 PageSlider 加入 pages。

9.1.5 编译运行

连接远程模拟器，编译运行，结果如图 9-4 所示，"我的"页签显示个人设置界面。

9.1.6 提交代码到仓库

将此功能的代码提交到本地仓库，并添加日志"创建我的界面"。

图 9-4 我的页签

9.2 云平台参数设置界面

在"我的"界面单击"云平台参数设置"，跳转到参数设置界面，用户可以在参数设置界面配置云平台相关参数，并单击"提交"按钮进入按钮单击事件处理流程。

9.2.1 创建云平台参数设置界面

1. 布局设计

在 layout 目录下新建 ability_cloud_parameter_setting.xml。根据图 9-2 所示，充分利用嵌套布局的设计方法进行布局设计。完整代码可以参考本书配套代码，在代码目录中通过以下命令提取任务 9 的代码版本。

```
git checkout task9
```

2. 设置按钮单击效果

1）在 graphic 目录下新建 setting_btn_bg.xml 文件，代码如下。

```
1.  <?xml version="1.0" encoding="UTF-8" ?>
2.  <state-container
3.      xmlns:ohos="http://schemas.huawei.com/res/ohos">
4.
5.      <item
6.          ohos:element="$graphic:setting_btn_bg_pressed"
7.          ohos:state="component_state_pressed"/>
8.
9.      <item
```

```
10.        ohos:element="$graphic:setting_btn_bg_empty"
11.        ohos:state="component_state_empty"/>
12.
13. </state-container>
```

第 5～7 行，设置按钮按下时的背景。

第 9～11 行，设置按钮未按下时的背景。

2）在 graphic 目录下新建 setting_btn_bg_pressed.xml，代码如下。

```
1.  <?xml version="1.0" encoding="UTF-8" ?>
2.  <state-container
3.      xmlns:ohos="http://schemas.huawei.com/res/ohos">
4.
5.      <item
6.          ohos:element="$graphic:setting_btn_bg_pressed"
7.          ohos:state="component_state_pressed"/>
8.
9.      <item
10.         ohos:element="$graphic:setting_btn_bg_empty"
11.         ohos:state="component_state_empty"/>
12.
13. </state-container>
```

3）在 graphic 目录下新建 setting_btn_bg_empty.xml，代码如下。

```
1.  <?xml version="1.0" encoding="UTF-8" ?>
2.  <shape
3.      xmlns:ohos="http://schemas.huawei.com/res/ohos"
4.      ohos:shape="rectangle">
5.
6.      <solid
7.          ohos:color="#FF33DDEC"/>
8.
9.      <corners
10.         ohos:radius="30"/>
11. </shape>
```

3．引用字符串

在 base /element 目录下，更新 string.json，代码如下。

```
1.  {
2.      "name": "strTxtDeviceId",
3.      "value": "设备 ID:"
4.  },
5.  {
6.      "name": "strTxtAirSetting",
7.      "value": "大气环境传感器参数设置"
8.  },
9.  {
10.     "name": "strTxtTemperatureId",
11.     "value": "温度 ID:"
12. },
13. {
```

```
14.        "name": "strTxtHumidityId",
15.        "value": "湿度 ID:"
16.    },
17.    {
18.        "name": "strTxtWindSpeedId",
19.        "value": "风速 ID:"
20.    },
21.    {
22.        "name": "strTxtWindDirectionId",
23.        "value": "风向 ID:"
24.    },
25.    {
26.        "name": "strTxtIlluminationId",
27.        "value": "光照 ID:"
28.    },
29.    {
30.        "name": "strTxtAtmosId",
31.        "value": "气压 ID:"
32.    },
33.    {
34.        "name": "strTxtPM2_5Id",
35.        "value": "PM2.5ID:"
36.    },
37.    {
38.        "name": "strTxtCarbonDioxideId",
39.        "value": "CO2ID:"
40.    },
41.    {
42.        "name": "strTxtSoilSetting",
43.        "value": "土壤环境传感器参数设置"
44.    },
45.    {
46.        "name": "strTxtPHId",
47.        "value": "PHID:"
48.    },
49.    {
50.        "name": "strTxtRainfallId",
51.        "value": "雨量 ID:"
52.    },
53.    {
54.        "name": "strTxtSoilTemperatureId",
55.        "value": "温度 ID:"
56.    },
57.    {
58.        "name": "strTxtSoilHumidityId",
59.        "value": "湿度 ID:"
60.    },
61.    {
62.        "name": "strTxtControlSetting",
63.        "value": "执行器传感器参数设置"
64.    },
```

```
65.    {
66.        "name": "strTxtWaterValve1Id",
67.        "value": "水阀1ID:"
68.    },
69.    {
70.        "name": "strTxtWaterValve2Id",
71.        "value": "水阀2ID:"
72.    },
73.    {
74.        "name": "strTxtWaterValve3Id",
75.        "value": "水阀3ID:"
76.    },
77.    {
78.        "name": "strTxtWaterValve4Id",
79.        "value": "水阀4ID:"
80.    },
81.    {
82.        "name": "strBtnSetting",
83.        "value": "保存参数"
84.    }
```

9.2.2 创建 AbilitySlice 的 Java 文件

在 slice 目录下新建 CloudParameterSettingAbilitySlice.java，完整代码如下。

```
1.  package com.example.smartagriculture.slice;
2.
3.  import com.example.smartagriculture.ResourceTable;
4.  import ohos.aafwk.ability.AbilitySlice;
5.  import ohos.aafwk.content.Intent;
6.  import ohos.agp.components.Button;
7.  import ohos.agp.components.Component;
8.
9.  public class CloudParameterSettingAbilitySlice extends AbilitySlice {
10.
11.     private Button btnSetting;
12.
13.     @Override
14.     protected void onStart(Intent intent) {
15.         super.onStart(intent);
16.         super.setUIContent(ResourceTable.Layout_ability_cloud_parameter_setting);
17.         initComponent();
18.         initClickedListener();
19.     }
20.
21.     private void initClickedListener() {
22.         btnSetting.setClickedListener(new Component.ClickedListener() {
23.             @Override
24.             public void onClick(Component component) {
```

```
25.                    System.out.println("云平台参数保存成功");
26.                    //销毁当前的 Ability Slice
27.                    terminate();
28.                }
29.            });
30.        }
31.
32.        private void initComponent() {
33.            btnSetting = (Button) findComponentById(ResourceTable.Id_btnSetting);
34.        }
35.  }
```

第 17 行，调用 initComponent 方法初始化组件。

第 18 行，调用 initClickedListener 方法初始化组件单击事件处理。

第 21～30 行，实现按钮单击事件处理，此处打印日志，并且通过 terminate 结束当前页面的生命周期。

第 32～34 行，initComponent 方法的具体实现，过程中初始化按钮组件。

9.2.3 更新 MainAbilitySlice.java

1. 增加成员属性

```
private Text txtCloudParameterSetting;
```

2. 修改 onStart 方法

```
1.  @Override
2.  public void onStart(Intent intent) {
3.      super.onStart(intent);
4.      super.setUIContent(ResourceTable.Layout_ability_main);
5.      initTabList();
6.      initPageSlider();
7.      initComponent();
8.      initClickedListener();
9.  }
```

第 7 行，initComponent 方法实现如下。

```
private void initComponent() {
    txtCloudParameterSetting = (Text) directionalLayoutPagesliderMe.
findComponentById(ResourceTable.Id_txtCloudParameterSetting);
}
```

在"我的"PageSlider 页面里获取"云平台参数设置"组件。

第 8 行，initClickedListener 方法实现如下。

```
private void initClickedListener() {
    txtCloudParameterSetting.setClickedListener(this);
}
```

setClickedListener 方法中传入 this 对象，实现组件的单击事件处理。该方法是通过当前类 MainAbilitySlice 实现 Component.ClickedListener 接口，即

```
    public class MainAbilitySlice extends AbilitySlice implements Component.
ClickedListener
```

3. 实现 Component.ClickedListener 接口

要实现 Component.ClickedListener 接口，则必须实现 onClick 方法，代码如下。

```
1.    @Override
2.    public void onClick(Component component) {
3.        int componentId = component.getId();
4.        switch (componentId) {
5.            case ResourceTable.Id_txtCloudParameterSetting: {
6.                System.out.println("MainAbilitySlice 切换到 CloudParameter-
SettingAbilitySlice");
7.                Intent intent = new Intent();
8.                present(new CloudParameterSettingAbilitySlice(), intent);
9.                break;
10.           }
11.       }
12.   }
```

第 3 行，通过 component.getId()获取单击的组件的 id。

第 5～9 行，判断当前单击的组件的 id 是不是 ResourceTable.Id_txtCloudParameterSetting，然后进入具体单击事件处理流程，此处是打印日志，并实现页面跳转，跳转到参数设置界面。

9.2.4 编译运行

连接远程模拟器，编译运行，运行结果如图 9-2 所示，单击"保存参数"，云平台参数设置界面关闭。

9.3 提交代码到仓库

将任务 9 的代码提交到本地仓库，完整流程如下。

1）通过"git status"查看当前目录状态。
2）通过"git add ."将当前目录下所有改动加入暂存区。
3）通过"git commit -m 创建参数设置界面"，提交暂存区内容到本地版本库。
4）通过"git status"查看当前工作目录，可以看到目录干净。
5）通过"git tag -a task9 -m 创建参数设置界面"，创建任务 9 标签。
6）通过"git log --pretty=oneline"查看版本库日志，可以看到提交的最新内容，以及 task9 标签，即任务 9 的版本。

任务 10　参数持久化

任务概述

任务 9 已经完成参数设置界面，本任务将完成数据的存储，App 将不用每次输入账号和云平台的参数，只需一次配置，就会保存到设备里，下次重启 App，就可以从设备里读数据。

知识目标

- 掌握 AbilityPackage 子类的使用。
- 掌握 Preferences 的使用。
- 掌握轻量级数据存储。

能力目标

- 能使用 AbilityPackage 子类存储公共成员属性。
- 能使用 Preferences 保存数据。

10.1　AbilityPackage 类

HarmonyOS 启动 App 时会自动创建 AbilityPackage 对象，一般用于初始化 App，存储系统的一些信息。我们可以通过简单创建 AbilityPackage 的子类，用以存储 App 的公共数据，初始化一些公共对象等。

AbilityPackage 是单例（singleton）模式，HarmonyOS 系统会在每个程序运行时创建一个 AbilityPackage 类的对象且仅创建一个，且 AbilityPackage 对象的生命周期是整个程序中最长的，它的生命周期就等于该程序的生命周期。因为它是全局的、单例的，在不同的 Ability 中获得的对象都是同一个对象，所以可以通过 AbilityPackage 来进行数据传递、数据共享、数据缓存等操作。

10.2　轻量级数据存储

对 App 的配置参数，云平台参数、账号等的存取可以通过读写文件、数据库和轻量级数据存储等方式实现，其中轻量级数据存储可以很方便地实现对上述数据的存取。

10.2.1　轻量级数据存储概述

1. 基本概念

轻量级数据存储适用于对 Key-Value 结构的数据进行存取和持久化操作。应用运行时全量数据将会被加载在内存中，使得访问速度更快，存取效率更高。如果对数据持久化，数据最终会写到文本文件中,建议在开发过程中减少写的频率,即减少对持久化文件的写次数。

轻量级数据存储是采用 Key-Value 数据结构的非关系型数据库。

(1) Key-Value 数据结构

一种键值结构数据类型。Key 是不重复的关键字,Value 是数据值。

(2) 非关系型数据库

区别于关系数据库,不保证遵循 ACID(Atomic、Consistency、Isolation 及 Durability)特性,不采用关系模型来组织数据,数据之间无关系,扩展性好。

2. 运作机制

1)本模块提供轻量级数据存储的操作类,应用通过这些操作类完成数据库操作。

2)借助轻量级数据存储的 API,应用可以将指定文件的内容加载到 Preferences 实例,每个文件最多有一个 Preferences 实例,系统会通过静态容器将该实例存储在内存中,直到应用主动从内存中移除该实例或者删除该文件。

3)获取文件对应的 Preferences 实例后,应用可以借助 Preferences API,从 Preferences 实例中读取数据或者将数据写入 Preferences 实例,通过 flush 或者 flushSync 将 Preferences 实例持久化,如图 10-1 所示。

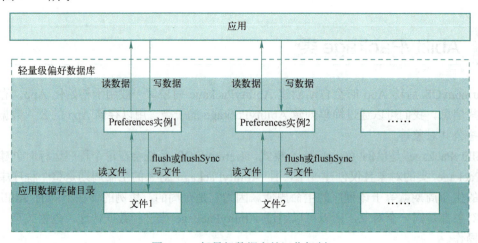

图 10-1 轻量级数据存储运作机制

3. 约束与限制

1)Key 键为 String 类型,要求非空且长度不超过 80 个字符。

2)Value 值为 String 类型时,可以为空但是长度不超过 8192 个字符。

3)Value 值为字符串型 Set 集合类型时,要求集合元素非空且长度不超过 8192 个字符。

4)存储的数据量应该是轻量级的,建议存储的数据不超过一万条,否则会在内存方面产生较大的开销。

10.2.2 轻量级数据存储开发

1. 场景介绍

轻量级数据存储主要用于保存应用的一些常用配置,并不适合存储大量数据和频繁改变数据的场景。用户的数据保存在文件中，可以持久存储在设备上。需要注意的是用户访问的实例包含文件的所有数据，这些数据一直加载在设备的内存中，并通过轻量级数据存储的 API 完成数据操作。

2. 接口说明

轻量级数据存储向本地应用提供的 API 支持本地应用读写数据及观察数据变化。数据存储形式为键值对，键的类型为字符串型，值的存储数据类型包括整型、字符串型、布尔型、浮点型、长整型、字符串型 Set 集合。

（1）创建数据库

通过数据库操作的辅助类可以获取要操作的 Preferences 实例，用于进行数据库的操作。轻量级偏好数据库创建接口如表 10-1 所示。

表 10-1　轻量级偏好数据库创建接口

类名	接口名	描述
DatabaseHelper	DatabaseHelper (Context context)	DatabaseHelper 是数据库操作的辅助类，当数据库创建成功后，数据库文件将存储在由上下文指定的目录里。数据库文件存储的路径会因指定不同的上下文而存在差异 获取上下文参考方法：ohos.app.Context#getApplicationContext()、ohos.app.AbilityContext#getContext() 查看详细路径信息：ohos.app.Context#getDatabaseDir()
	Preferences getPreferences(String name)	获取文件对应的 Preferences 单实例，用于数据操作

（2）查询数据

通过调用 get 系列的方法，可以查询不同类型的数据，如表 10-2 所示。

表 10-2　轻量级偏好数据库查询接口

类名	接口名	描述
Preferences	int getInt(String key, int defValue)	获取键对应的 int 类型的值
Preferences	float getFloat(String key, float defValue)	获取键对应的 float 类型的值

（3）插入数据

通过 put 系列的方法，可以修改 Preferences 实例中的数据，通过 flush 或者 flushSync 将 Preferences 实例持久化，如表 10-3 所示。

表 10-3　轻量级偏好数据库插入接口

类名	接口名	描述
Preferences	Preferences putInt(String key, int value)	设置 Preferences 实例中键对应的 int 类型的值
Preferences	Preferences putString(String key, String value)	设置 Preferences 实例中键对应的 String 类型的值
Preferences	void flush()	将 Preferences 实例异步写入文件
Preferences	boolean flushSync()	将 Preferences 实例同步写入文件

（4）观察数据变化

轻量级数据存储还提供了一系列的接口变化回调，用于应用响应数据的变化。开发者可以通过重写 onChange 方法来定义观察者的行为，如表 10-4 所示。

表 10-4　轻量级数据存储接口变化回调

类名	接口名	描述
Preferences	void registerObserver (PreferencesObserver preferencesObserver)	注册观察者，用于观察数据变化

（续）

类名	接口名	描述
Preferences	void unRegisterObserver(PreferencesObserver preferencesObserver)	注销观察者
Preferences.PreferencesObserver	void onChange(Preferences preferences, String key)	观察者的回调方法，任意数据变化都会回调该方法

（5）删除数据文件

通过调用以下两种接口，可以删除数据文件，如表 10-5 所示。

表 10-5　轻量级数据存储删除接口

类名	接口名	描述
DatabaseHelper	boolean deletePreferences(String name)	删除文件和文件对应的 Preferences 实例
DatabaseHelper	void removePreferencesFromCache(String name)	删除文件对应的 Preferences 实例

（6）移动数据库文件

轻量级数据存储移动接口如表 10-6 所示。

表 10-6　轻量级数据存储移动接口

类名	接口名	描述
DatabaseHelper	boolean movePreferences(Context sourceContext, String sourceName, String targetName)	移动数据库文件

3. 开发步骤

（1）准备工作

导入轻量级偏好数据库包到开发环境。

（2）获取 Preferences 实例

读取指定文件，将数据加载到 Preferences 实例，用于数据操作。

```
1. Context context = getContext(); // 数据文件存储路径：/data/data/{PackageName}/{AbilityName}/preferences。
2. // Context context = getApplicationContext(); // 数据文件存储路径：/data/data/{PackageName}/preferences。
3. DatabaseHelper databaseHelper = new DatabaseHelper(context); // context 入参类型为 ohos.app.Context。
4. String fileName = "test_pref"; // fileName 表示文件名，其取值不能为空，也不能包含路径，默认存储目录可以通过 context.getPreferencesDir()获取。
5. Preferences preferences = databaseHelper.getPreferences(fileName);
```

（3）从指定文件读取数据

首先获取指定文件对应的 Preferences 实例，然后借助 Preferences API 读取数据。读取整型数据示例如下。

```
int value = preferences.getInt("intKey", 0);
```

（4）将数据写入指定文件

首先获取指定文件对应的 Preferences 实例，然后借助 Preferences API 将数据写入 Preferences 实例，通过 flush 或者 flushSync 将 Preferences 实例持久化。

异步（flush）：

```
1. preferences.putInt("intKey", 3);
2. preferences.putString("StringKey", "String value");
3. preferences.flush();
```

同步（flushSync）：

```
1. preferences.putInt("intKey", 3);
2. preferences.putString("StringKey", "String value");
3. bool result = preferences.flushSync();
```

（5）注册观察者。

开发者可以向 Preferences 实例注册观察者，观察者对象需实现 Preferences.PreferencesObserver 接口。flushSync()或 flush()执行后，该 Preferences 实例注册的所有观察者的 onChange()方法都会被回调。不再需要观察者时请注销。

```
1.  private class PreferencesObserverImpl implements Preferences.PreferencesObserver {
2.
3.      @Override
4.      public void onChange(Preferences preferences, String key) {
5.          if ("intKey".equals(key)) {
6.              HiLog.info(LABLE, "Change Received：[key=value]");
7.          }
8.      }
9.  }
10.
11. // 向 preferences 实例注册观察者
12. PreferencesObserverImpl observer = new PreferencesObserverImpl();
13. preferences.registerObserver(observer);
14. // 修改数据
15. preferences.putInt("intKey", 3);
16. preferences.flush();
17. // 修改数据后，observer 的 onChange 方法会被回调
18. // 向 preferences 实例注销观察者
19. preferences.unRegisterObserver(observer);
```

（6）移除 Preferences 实例

从内存中移除指定文件对应的 Preferences 单实例。移除 Preferences 单实例时，应用不允许再使用该实例进行数据操作，否则会出现数据一致性问题。

```
1. DatabaseHelper databaseHelper = new DatabaseHelper(context);
2. String fileName = "name"; // fileName 表示文件名，其取值不能为空，也不能包含路径。
3. databaseHelper.removePreferencesFromCache(fileName);
```

（7）删除指定文件

从内存中移除指定文件对应的 Preferences 单实例，并删除指定文件及其备份文件、损坏文件。删除指定文件时，应用不允许再使用该实例进行数据操作，否则会出现数据一致性问题

```
1. DatabaseHelper databaseHelper = new DatabaseHelper(context);
2. String fileName = "name"; // fileName 表示文件名，其取值不能为空，也不能包
```
含路径。

```
3.    boolean result = databaseHelper.deletePreferences(fileName);
```

(8)移动指定文件

从源路径移动文件到目标路径。移动文件时,应用不允许再操作该文件数据,否则会出现数据一致性问题。

```
1.    Context targetContext = getContext();
2.    DatabaseHelper databaseHelper = new DatabaseHelper(targetContext);
3.    String srcFile = "srcFile"; // srcFile 表示源文件名或者源文件的绝对路径,不能为相对路径,其取值不能为空。当 srcFile 只传入文件名时,srcContext 不能为空。
4.    String targetFile = "targetFile"; // targetFile 表示目标文件名,其取值不能为空,也不能包含路径。
5.    Context srcContext = getApplicationContext();
6.    boolean result = databaseHelper.movePreferences(srcContext, srcFile, targetFile);
```

10.3 更新 Java 代码

智慧农业 App 的数据存储,通过获取参数界面的数据和登录界面的账号数据,存储到 AbilityPackage 子类对象的成员属性中,并通过轻量级数据存储实现参数的持久化。

10.3.1 更新 MyApplication.java 文件

修改 MyApplication.java,该类继承了 AbilityPackage,代码如下。

```
1.  package com.example.smartagriculture;
2.
3.  import com.example.smartagriculture.net.Wan;
4.  import ohos.aafwk.ability.AbilityPackage;
5.  import ohos.data.DatabaseHelper;
6.  import ohos.data.preferences.Preferences;
7.  import poerty.jianjian.converter.gson.GsonConverterFactory;
8.  import poetry.jianjia.JianJia;
9.
10. public class MyApplication extends AbilityPackage {
11.     private JianJia mJianJia;
12.     private Wan mWan;
13.     private Preferences preferences;
14.     private String PREFERENCES_FILE_CLOUD_PARAMETERS = "SmartAgriculture";
15.
16.     private String account, password, telephone, college, role, accesstoken;
17.     private boolean isLogin, isCloudParameterSetting;
18.     private String deviceId;
19.
20.     //大气环境传感器 ID 参数
21.     private String temperatureID, humidityID, windSpeedID,
```

```
   windDirectionID, illuminationID, atmosID, pm2_5ID, carbonDioxideID;
22.        //土壤环境传感器ID参数
23.        private String PHID, rainfallID, soilTemperatureID, soilHumidityID;
24.        //执行器传感器ID参数
25.        private String waterValve1ID, waterValve2ID, waterValve3ID, waterValve4ID;
26.
27.        @Override
28.        public void onInitialize() {
29.            super.onInitialize();
30.            mJianJia = new JianJia.Builder()
31.                    .baseUrl("http://api.nlecloud.com")
32.                    .addConverterFactory(GsonConverterFactory.create())
33.                    .build();
34.            mWan = mJianJia.create(Wan.class);
35.            DatabaseHelper databaseHelper = new DatabaseHelper(this);
36.            preferences = databaseHelper.getPreferences(PREFERENCES_
    FILE_CLOUD_PARAMETERS);
37.            readAccount();
38.            readParameters();
39.        }
40.
41.        public void readAccount() {
42.            isLogin = preferences.getBoolean("isLogin", false);
43.            if (isLogin) {
44.                account = preferences.getString("account", "");
45.                password = preferences.getString("Password", "");
46.                accesstoken = preferences.getString("AccessToken", "");
47.            }
48.        }
49.
50.        public void readParameters() {
51.            isCloudParameterSetting = preferences.getBoolean("isCloudPa-
    rameterSetting", false);
52.            if (isLogin && isCloudParameterSetting) {
53.                deviceId = preferences.getString("deviceId", "");
54.                temperatureID = preferences.getString("temperatureID", "");
55.                humidityID = preferences.getString("humidityID", "");
56.                windSpeedID = preferences.getString("windSpeedID", "");
57.                windDirectionID = preferences.getString("windDirectionID", "");
58.                illuminationID = preferences.getString("illuminationID", "");
59.                atmosID = preferences.getString("atmosID", "");
60.                pm2_5ID = preferences.getString("pm2_5ID", "");
61.                carbonDioxideID = preferences.getString("carbonDioxideID", "");
62.
63.                PHID = preferences.getString("PHID", "");
64.                rainfallID = preferences.getString("rainfallID", "");
65.                soilTemperatureID = preferences.getString("soilTemperat-
    ureID", "");
```

```java
66.              soilHumidityID = preferences.getString("soilHumidityID", "");
67.
68.              waterValve1ID = preferences.getString("waterValve1ID", "");
69.              waterValve2ID = preferences.getString("waterValve2ID", "");
70.              waterValve3ID = preferences.getString("waterValve3ID", "");
71.              waterValve4ID = preferences.getString("waterValve4ID", "");
72.          }
73.      }
74.
75.      public JianJia getmJianJia() {
76.          return mJianJia;
77.      }
78.
79.      public Wan getmWan() {
80.          return mWan;
81.      }
82.
83.      public Preferences getPreferences() {
84.          return preferences;
85.      }
86.
87.      public String getAccount() {
88.          return account;
89.      }
90.
91.      public void setAccount(String account) {
92.          this.account = account;
93.      }
94.
95.      public String getPassword() {
96.          return password;
97.      }
98.
99.      public void setPassword(String password) {
100.         this.password = password;
101.     }
102.
103.     public String getTelephone() {
104.         return telephone;
105.     }
106.
107.     public void setTelephone(String telephone) {
108.         this.telephone = telephone;
109.     }
110.
111.     public String getCollege() {
112.         return college;
113.     }
```

```
114.
115.        public void setCollege(String college) {
116.            this.college = college;
117.        }
118.
119.        public String getRole() {
120.            return role;
121.        }
122.
123.        public void setRole(String role) {
124.            this.role = role;
125.        }
126.
127.        public String getAccesstoken() {
128.            return accesstoken;
129.        }
130.
131.        public void setAccesstoken(String accesstoken) {
132.            this.accesstoken = accesstoken;
133.        }
134.
135.        public boolean isLogin() {
136.            return isLogin;
137.        }
138.
139.        public void setLogin(boolean login) {
140.            isLogin = login;
141.        }
142.
143.        public boolean isCloudParameterSetting() {
144.            return isCloudParameterSetting;
145.        }
146.
147.        public void setCloudParameterSetting(boolean cloudParameter-
Setting) {
148.            isCloudParameterSetting = cloudParameterSetting;
149.        }
150.
151.        public String getDeviceId() {
152.            return deviceId;
153.        }
154.
155.        public void setDeviceId(String deviceId) {
156.            this.deviceId = deviceId;
157.        }
158.
159.        public String getTemperatureID() {
```

```
160.            return temperatureID;
161.        }
162.
163.        public void setTemperatureID(String temperatureID) {
164.            this.temperatureID = temperatureID;
165.        }
166.
167.        public String getHumidityID() {
168.            return humidityID;
169.        }
170.
171.        public void setHumidityID(String humidityID) {
172.            this.humidityID = humidityID;
173.        }
174.
175.        public String getWindSpeedID() {
176.            return windSpeedID;
177.        }
178.
179.        public void setWindSpeedID(String windSpeedID) {
180.            this.windSpeedID = windSpeedID;
181.        }
182.
183.        public String getWindDirectionID() {
184.            return windDirectionID;
185.        }
186.
187.        public void setWindDirectionID(String windDirectionID) {
188.            this.windDirectionID = windDirectionID;
189.        }
190.
191.        public String getIlluminationID() {
192.            return illuminationID;
193.        }
194.
195.        public void setIlluminationID(String illuminationID) {
196.            this.illuminationID = illuminationID;
197.        }
198.
199.        public String getAtmosID() {
200.            return atmosID;
201.        }
202.
203.        public void setAtmosID(String atmosID) {
204.            this.atmosID = atmosID;
205.        }
206.
```

```java
207.        public String getPm2_5ID() {
208.            return pm2_5ID;
209.        }
210.
211.        public void setPm2_5ID(String pm2_5ID) {
212.            this.pm2_5ID = pm2_5ID;
213.        }
214.
215.        public String getCarbonDioxideID() {
216.            return carbonDioxideID;
217.        }
218.
219.        public void setCarbonDioxideID(String carbonDioxideID) {
220.            this.carbonDioxideID = carbonDioxideID;
221.        }
222.
223.        public String getPHID() {
224.            return PHID;
225.        }
226.
227.        public void setPHID(String PHID) {
228.            this.PHID = PHID;
229.        }
230.
231.        public String getRainfallID() {
232.            return rainfallID;
233.        }
234.
235.        public void setRainfallID(String rainfallID) {
236.            this.rainfallID = rainfallID;
237.        }
238.
239.        public String getSoilTemperatureID() {
240.            return soilTemperatureID;
241.        }
242.
243.        public void setSoilTemperatureID(String soilTemperatureID) {
244.            this.soilTemperatureID = soilTemperatureID;
245.        }
246.
247.        public String getSoilHumidityID() {
248.            return soilHumidityID;
249.        }
250.
251.        public void setSoilHumidityID(String soilHumidityID) {
252.            this.soilHumidityID = soilHumidityID;
253.        }
```

```
254.
255.        public String getWaterValve1ID() {
256.            return waterValve1ID;
257.        }
258.
259.        public void setWaterValve1ID(String waterValve1ID) {
260.            this.waterValve1ID = waterValve1ID;
261.        }
262.
263.        public String getWaterValve2ID() {
264.            return waterValve2ID;
265.        }
266.
267.        public void setWaterValve2ID(String waterValve2ID) {
268.            this.waterValve2ID = waterValve2ID;
269.        }
270.
271.        public String getWaterValve3ID() {
272.            return waterValve3ID;
273.        }
274.
275.        public void setWaterValve3ID(String waterValve3ID) {
276.            this.waterValve3ID = waterValve3ID;
277.        }
278.
279.        public String getWaterValve4ID() {
280.            return waterValve4ID;
281.        }
282.
283.        public void setWaterValve4ID(String waterValve4ID) {
284.            this.waterValve4ID = waterValve4ID;
285.        }
286.    }
```

第 11、12 行，增加蒹葭的成员属性。

第 13、14 行，增加 Preferences 相关成员属性。

第 16 行，增加账号相关成员属性。

第 17 行，增加是否登录和参数是否设置的成员属性。

第 18～25 行，增加云平台参数的成员属性。

第 30～34 行，初始化蒹葭网络库相关对象。

第 35、36 行，创建轻量级数据存储对象。

第 37 行，调用 readAccount 方法，其实现位于第 41～48 行。

第 41～48 行，readAccount 方法的实现，首先读取持久化文件里的 isLogin，判断是否登录，如果登录，则读取持久化文件里的账号信息和令牌数据。

第 38 行，调用 readParameters 方法。

第 50～73 行，readParameters 方法的实现，读取持久化文件的 isCloudParameterSetting，判

断用户是否已经配置了云平台参数，如果配置了，则读取云平台参数并赋值给 AbilityPackage 子类的成员属性。

剩余代码是 AbilityPackage 子类的成员属性的 Setter 与 Getter 方法。

10.3.2 更新 CloudParameterSettingAbilitySlice.java

1. 增加成员属性

```
1.   private TextField tfDeviceIdValue;
2.   //大气环境 TextField 组件
3.   private TextField tfTemperatureIdValue, tfHumidityIdValue,
4.           tfWindSpeedIdValue, tfWindDirectionIdValue,
5.           tfIlluminationIdValue, tfAtmosIdValue,
6.           tfPM2_5IdValue, tfCarbonDioxideIdValue;
7.   //土壤环境 TextField 组件
8.   private TextField tfPHIdValue, tfRainfallIdValue,
9.           tfSoilTemperatureIdValue, tfSoilHumidityIdValue;
10.  //水阀 TextField 组件
11.  private TextField tfWaterValve1IdValue, tfWaterValve2IdValue,
12.          tfWaterValve3IdValue, tfWaterValve4IdValue;
13.  //设备 ID 参数
14.  private String deviceId;
15.  //大气环境传感器 ID 参数
16.  private String temperatureID, humidityID, windSpeedID, windDirectionID, illuminationID, atmosID, pm2_5ID, carbonDioxideID;
17.  //土壤环境传感器 ID 参数
18.  private String PHID, rainfallID, soilTemperatureID, soilHumidityID;
19.  //执行器传感器 ID 参数
20.  private String waterValve1ID, waterValve2ID, waterValve3ID, waterValve4ID;
21.  private MyApplication application;
22.  private Preferences preferences;
```

2. 修改 onStart 方法

在 onStart 中，增加如下成员属性的初始化。

```
application = (MyApplication) getAbility().getAbilityPackage();
preferences = application.getPreferences();
```

获取 MyApplication 对象和 Preferences 对象。

3. 修改 initClickedListener 方法

```
1.   private void initClickedListener() {
2.       btnSetting.setClickedListener(new Component.ClickedListener() {
3.           @Override
4.           public void onClick(Component component) {
5.               System.out.println("云平台参数保存成功");
6.               ToastDialog toastDialog = new ToastDialog(getContext()).setAlignment(LayoutAlignment.CENTER);
7.               if (!application.isLogin()) {
```

```
8.                    toastDialog.setText("请先登录")
9.                            .show();
10.                   return;
11.           }
12.           getParameters();
13.           if (isNull()) {
14.                   toastDialog.setText("参数不能为空")
15.                           .show();
16.                   return;
17.           }
18.           writeParameters();
19.           application.readParameters();
20.           toastDialog.setText("参数保存成功！")
21.                   .show();
22.           System.out.println("IlluminationID: " + application.
getIlluminationID());
23.           //销毁当前的 Ability Slice
24.           terminate();
25.       }
26.   });
27. }
```

onClick 单击事件处理流程图如图 10-2 所示，首先判断用户是否登录，如果没有登录，则弹出对话框，提醒用户先登录，如果登录了，先获取界面参数。然后判断参数是否为空，如果为空，则弹出对话框，提醒用户参数不能为空，如果参数不为空，则将参数写入持久化文件。等持久化完成后，再提取数据并赋值给 MyApplication 对象的成员属性。最后弹出对话框，告诉用户参数保存成功，并关闭当前界面。

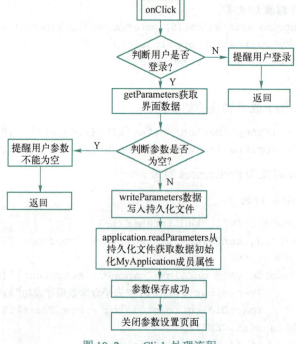

图 10-2　onClick 处理流程

4. 修改 initComponent 方法

```
1.  private void initComponent() {
2.      btnSetting = (Button) findComponentById(ResourceTable.Id_
btnSetting);
3.      //获取设备组件
4.      tfDeviceIdValue = (TextField) findComponentById(ResourceTable.
Id_tfDeviceIdValue);
5.      //获取大气环境组件
6.      tfTemperatureIdValue = (TextField) findComponentById(Resource-
Table.Id_tfTemperatureIdValue);
7.      tfHumidityIdValue = (TextField) findComponentById(ResourceTable.
Id_tfHumidityIdValue);
8.      tfWindSpeedIdValue = (TextField) findComponentById(ResourceTable.
Id_tfWindSpeedIdValue);
9.      tfWindDirectionIdValue = (TextField) findComponentById(Resource-
Table.Id_tfWindDirectionIdValue);
10.     tfIlluminationIdValue = (TextField) findComponentById(Resource-
Table.Id_tfIlluminationIdValue);
11.     tfAtmosIdValue = (TextField) findComponentById(ResourceTable.Id_
tfAtmosIdValue);
12.     tfPM2_5IdValue = (TextField) findComponentById(ResourceTable.Id_
tfPM2_5IdValue);
13.     tfCarbonDioxideIdValue = (TextField) findComponentById(Resource-
Table.Id_tfCarbonDioxideIdValue);
14.     //获取土壤环境组件
15.     tfPHIdValue = (TextField) findComponentById(ResourceTable.Id_
tfPHIdValue);
16.     tfRainfallIdValue = (TextField) findComponentById(ResourceTable.
Id_tfRainfallIdValue);
17.     tfSoilTemperatureIdValue = (TextField) findComponentById(Resour-
ceTable.Id_tfSoilTemperatureIdValue);
18.     tfSoilHumidityIdValue = (TextField) findComponentById(ResourceT-
able.Id_tfSoilHumidityIdValue);
19.     //获取水阀组件
20.     tfWaterValve1IdValue = (TextField) findComponentById(ResourceTa-
ble.Id_tfWaterValve1IdValue);
21.     tfWaterValve2IdValue = (TextField) findComponentById(ResourceTa-
ble.Id_tfWaterValve2IdValue);
22.     tfWaterValve3IdValue = (TextField) findComponentById(ResourceTa-
ble.Id_tfWaterValve3IdValue);
23.     tfWaterValve4IdValue = (TextField) findComponentById(ResourceTa-
ble.Id_tfWaterValve4IdValue);
24. }
```

获取参数设置界面的云平台参数的组件，并赋值给相对应的成员属性。

5. 增加 getParameters、isNull 和 writeParameters 方法的实现

```
1.  private void getParameters() {
```

```
2.        deviceId = tfDeviceIdValue.getText().trim();
3.        temperatureID = tfTemperatureIdValue.getText().trim();
4.        humidityID = tfHumidityIdValue.getText().trim();
5.        windSpeedID = tfWindSpeedIdValue.getText().trim();
6.        windDirectionID = tfWindDirectionIdValue.getText().trim();
7.        illuminationID = tfIlluminationIdValue.getText().trim();
8.        atmosID = tfAtmosIdValue.getText().trim();
9.        pm2_5ID = tfPM2_5IdValue.getText().trim();
10.       carbonDioxideID = tfCarbonDioxideIdValue.getText().trim();
11.       PHID = tfPHIdValue.getText().trim();
12.       rainfallID = tfRainfallIdValue.getText().trim();
13.       soilTemperatureID = tfSoilTemperatureIdValue.getText().trim();
14.       soilHumidityID = tfSoilHumidityIdValue.getText().trim();
15.       waterValve1ID = tfWaterValve1IdValue.getText().trim();
16.       waterValve2ID = tfWaterValve2IdValue.getText().trim();
17.       waterValve3ID = tfWaterValve3IdValue.getText().trim();
18.       waterValve4ID = tfWaterValve4IdValue.getText().trim();
19. }
20. public boolean isNull() {
21.       if (deviceId.equals("") || temperatureID.equals("") || humidityID.equals("") || windSpeedID.equals("")
22.              || windDirectionID.equals("") || illuminationID.equals("") || atmosID.equals("")
23.              || pm2_5ID.equals("") || carbonDioxideID.equals("") || PHID.equals("") || rainfallID.equals("")
24.              || soilTemperatureID.equals("") || soilHumidityID.equals("") || waterValve1ID.equals("")
25.              || waterValve2ID.equals("") || waterValve3ID.equals("") || waterValve4ID.equals("")) {
26.            return true;
27.       }
28.       return false;
29. }
30. public void writeParameters() {
31.       preferences.putString("deviceId", deviceId);
32.       preferences.putString("temperatureID", temperatureID);
33.       preferences.putString("humidityID", humidityID);
34.       preferences.putString("windSpeedID", windSpeedID);
35.       preferences.putString("windDirectionID", windDirectionID);
36.       preferences.putString("illuminationID", illuminationID);
37.       preferences.putString("atmosID", atmosID);
38.       preferences.putString("pm2_5ID", pm2_5ID);
39.       preferences.putString("carbonDioxideID", carbonDioxideID);
40.       preferences.putString("PHID", PHID);
41.       preferences.putString("rainfallID", rainfallID);
42.       preferences.putString("soilTemperatureID", soilTemperatureID);
43.       preferences.putString("soilHumidityID", soilHumidityID);
```

```
44.         preferences.putString("waterValve1ID", waterValve1ID);
45.         preferences.putString("waterValve2ID", waterValve2ID);
46.         preferences.putString("waterValve3ID", waterValve3ID);
47.         preferences.putString("waterValve4ID", waterValve4ID);
48.         //记录参数已设置标志
49.         preferences.putBoolean("isCloudParameterSetting", true);
50.         preferences.flushSync();
51.     }
```

10.3.3 更新 SplashAbilitySlice.java

1. 增加成员属性

```
private MyApplication application;
```

2. 修改 onStart 方法

在 onStart 中，增加如下成员属性的初始化。

```
application = (MyApplication) getAbility().getAbilityPackage();
```

3. 修改 goToLogin 方法

```
1.  private void goToLogin() {
2.      if (application.isLogin()) {
3.          Intent intent = new Intent();
4.          // 指定待启动 FA 的 bundleName 和 abilityName
5.          Operation operation = new Intent.OperationBuilder()
6.                  .withDeviceId("")
7.                  .withBundleName("com.example.smartagriculture")
8.                  .withAbilityName("com.example.smartagriculture.MainAbility")
9.                  .build();
10.         intent.setOperation(operation);
11.         startAbility(intent);
12.     } else {
13.         Intent intent = new Intent();
14.         present(new LoginAbilitySlice(), intent);
15.     }
16. }
```

判断用户是否已经登录，如果已经登录，则直接跳转主界面，否则跳转到登录界面。

10.3.4 更新 LoginAbilitySlice.java

1. 修改成员属性

去除成员属性 mJianJia 和 mWan，增加如下属性。

```
private MyApplication application;
private Preferences preferences;
```

2. 修改 onStart 方法

在 onStart 中，增加如下成员属性的初始化。

```
application = (MyApplication) getAbility().getAbilityPackage();
preferences = application.getPreferences();
```

去除成员属性 mJianJia 和 mWan 的赋值，兼葭网络库的对象通过 application 对象获取。

3. 修改 login 方法

```
1.   private void login(String account, String passwd) {
2.       System.out.println(TAG + "账号: " + account + "\n" + "密码: " + passwd);
3.       application.getmWan().login(account, passwd, true).enqueue(new Callback<Account>() {
4.           @Override
5.           public void onResponse(Call<Account> call, Response<Account> response) {
6.               try {
7.                   if(response.isSuccessful() && response.body().Status == 0){
8.                       System.out.println(TAG + "AccessToken: "+ response.body().ResultObj.AccessToken);
9.                       preferences.putString("account", account);
10.                      preferences.putString("Password", passwd);
11.                      preferences.putString("AccessToken", response.body().ResultObj.AccessToken);
12.                      preferences.putBoolean("isLogin", true);
13.                      preferences.flushSync();
14.                      application.readAccount();
15.                      goToMainAbility();
16.                  } else {
17.                      System.out.println(TAG + "登录出错" + response.body().Msg);
18.                  }
19.              } catch (Exception e) {
20.                  e.printStackTrace();
21.              }
22.          }
23.          @Override
24.          public void onFailure(Call<Account> call, Throwable throwable) {
25.              System.out.println(TAG + "登录请求出错"+ throwable.getMessage());
26.          }
27.      });
28.  }
```

如果登录认证成功，则将账号相关数据保存到持久化文件中，设置已经登录的标志，并跳转到主界面。

10.4 编译运行

连接远程模拟器,编译运行,运行结果如图 10-3 所示,单击"保存参数",云平台参数设置界面关闭,参数保存成功后,读取光照 ID 功能正常。

图 10-3　参数保存

10.5 提交代码到仓库

将任务 10 的代码提交到本地仓库,完整流程如下。

1)通过"git status"查看当前目录状态。
2)通过"git add."将当前目录下所有改动加入暂存区。
3)通过"git commit -m 实现参数持久化",提交暂存区内容到本地版本库。
4)通过"git status"查看当前工作目录,可以看到目录干净。
5)通过"git tag -a task10 -m 实现参数持久化",创建任务 10 标签。
6)通过"git log --pretty=oneline"查看版本库日志,可以看到提交的最新内容,以及 task10 标签,即任务 10 的版本。

任务 11　从云平台获取传感器数据

任务概述

如何在大气环境监控界面中显示传感器的实时数据？在任务 1 的系统概述中我们知道，传感器采集到数据后，通过无线或者有线的方式传送到物联网网关，然后物联网网关通过网络将数据传送到云平台，移动终端通过网络访问云平台获取传感器的实时数据并显示，如图 11-1 所示，任务 11 主要实现传感器数据的获取并显示。

图 11-1　大气环境监控界面

知识目标

- 掌握蒹葭（JianJia）网络库的使用。
- 掌握蒹葭（JianJia）拦截器打印日志的使用。
- 掌握 HiLog 日志打印。
- 掌握新大陆物联网云平台核心 RESTful API。

能力目标

- 能使用蒹葭（JianJia）从云平台获取数据。
- 能配置蒹葭（JianJia）拦截器打印网络请求响应的日志。
- 能使用 HiLog 调试程序。
- 能使用 Timer 定时更新传感器数据。

11.1　设置蒹葭（JianJia）拦截器

设置蒹葭拦截器，可以将网络请求与响应的过程打印出来，方便程序的调试。特别是当程序功能出现问题时，可以很方便地定位 App 与云平台之间的交互，快速找到出问题的地方。

11.1.1 蒹葭（JianJia）拦截器

使用拦截器打印日志，在 entry 目录下的 build.gradle 文件中的 dependencies 添加如下依赖，之前如果已经添加了，可以省略这一步。

```
implementation 'com.squareup.okhttp3:logging-interceptor:3.7.0'
```

更新了 build.gradle 文件后，需要单击软件右上角的"Sync Now"进行工程的同步操作。给 OkHttp（依赖库）添加拦截器，示例代码如下。

```
// 创建日志拦截器
HttpLoggingInterceptor logging = new HttpLoggingInterceptor();
logging.setLevel(HttpLoggingInterceptor.Level.BODY);
// 为 OkHttp 添加日志拦截器
OkHttpClient okHttpClient = new OkHttpClient.Builder()
    .addInterceptor(logging)
    .build();
// 创建全局的蒹葭对象
mJianJia = new JianJia.Builder()
    // 使用自定义的 okHttpClient 对象
    .callFactory(okHttpClient)
    .baseUrl("https://www.wanandroid.com")
    .addConverterFactory(GsonConverterFactory.create())
    .build();
mWan = mJianJia.create(Wan.class);
```

11.1.2 更新 MyApplication.java 文件

修改 MyApplication.java 文件中 onInitialize 方法，代码如下。

```
1.   public void onInitialize() {
2.       super.onInitialize();
3.       // 创建日志拦截器
4.       HttpLoggingInterceptor logging = new HttpLoggingInterceptor();
5.       logging.setLevel(HttpLoggingInterceptor.Level.BODY);
6.       // 为 OKHTTP 添加日志拦截器
7.       OkHttpClient okHttpClient = new OkHttpClient.Builder()
8.           .addInterceptor(logging)
9.           .build();
10.      mJianJia = new JianJia.Builder()
11.          // 使用自定义的 okHttpClient 对象
12.          .callFactory(okHttpClient)
13.          .baseUrl("http://api.nlecloud.com")
14.          .addConverterFactory(GsonConverterFactory.create())
15.          .build();
16.      mWan = mJianJia.create(Wan.class);
17.      DatabaseHelper databaseHelper = new DatabaseHelper(this);
18.      preferences = databaseHelper.getPreferences(PREFERENCES_FILE_CLOUD_PARAMETERS);
```

```
19.        readAccount();
20.        readParameters();
21.    }
```

创建 mJianJia 对象时，增加蒹葭日志拦截器。

11.1.3 编译运行

编译运行程序，在登录界面输入云平台的账号和密码，此时就会使用蒹葭（JianJia）网络库访问云平台验证账号，可以看到日志区域打印了请求和响应，如图 11-2 所示。虽然登录使用了 HTTP 的 POST 方法，使得账号密码等敏感信息不通过 URL 参数传递，包裹在请求体中，但因为是 HTTP，请求体不加密，也是明文传输，为了网络通信安全，就能理解为什么现在全球主推 HTTPS 进行网络访问，因为 HTTPS 会对请求体进行加密。

图 11-2　登录请求和响应日志

11.1.4 提交代码到仓库

将此功能的代码提交到本地仓库，并添加日志"增加蒹葭（JianJia）日志拦截器"。

11.2　使用 HiLog 日志

任务 11 之前，日志打印使用的是 System.out.println，这种方式的缺点是不能控制日志的打印，如果程序发布，必须删除这些日志打印程序，否则将影响程序性能。HiLog 提供了精细的日志打印功能和控制功能，解决了以上问题。

11.2.1 HiLog 日志基础

HarmonyOS 提供了 HiLog 日志系统,让应用可以按照指定类型、指定级别、指定格式字符串输出日志内容,帮助开发者了解应用的运行状态,更好地调试程序。

输出日志的接口由 HiLog 类提供。在输出日志前,需要先调用 HiLog 的辅助类 HiLogLabel 定义日志标签。

1. 定义日志标签

使用 HiLogLabel(int type, int domain, String tag)定义日志标签,其中包括了日志类型、业务领域和标签。使用示例如下。

```
static final HiLogLabel LABEL = new HiLogLabel(HiLog.LOG_APP, 0x00201, "MY_TAG");
```

1)参数 type:用于指定输出日志的类型。HiLog 中当前只提供了一种日志类型,即应用日志类型 LOG_APP。

2)参数 domain:用于指定输出日志所对应的业务领域,取值范围为 0x0~0xFFFFF,开发者可以根据需要进行自定义。一般情况下,建议把这 5 个十六进制数分成两组,前面三个数表示应用中的模块编号,后面两个表示模块中的类的编号。

3)参数 tag:用于指定日志标识,可以为任意字符串,建议标识调用所在的类或者业务行为。一般情况下将类的名字作为标识,这样可用这个标记对日志进行过滤。

开发者可以根据自定义参数 domain 和 tag 来进行日志的筛选和查找。

2. 输出日志

HiLog 中定义了 DEBUG、INFO、WARN、ERROR、FATAL 五种日志级别,并提供了对应的方法用于输出不同级别的日志,如表 11-1 所示。

表 11-1 HiLog 提供的主要接口

接口名	功能描述
debug(HiLogLabel label, String format, Object... args)	输出 DEBUG 级别的日志。DEBUG 级别日志表示仅用于应用调试,默认不输出,输出前需要在设备的"开发人员选项"中打开"USB 调试"开关
info(HiLogLabel label, String format, Object... args)	输出 INFO 级别的日志。INFO 级别日志表示普通的信息
warn(HiLogLabel label, String format, Object... args)	输出 WARN 级别的日志。WARN 级别日志表示存在警告
error(HiLogLabel label, String format, Object... args)	输出 ERROR 级别的日志。ERROR 级别日志表示存在错误
fatal(HiLogLabel label, String format, Object... args)	输出 FATAL 级别的日志。FATAL 级别日志表示出现致命错误、不可恢复错误

1)参数 label:定义好的 HiLogLabel 标签。

2)参数 format:格式字符串,用于日志的格式化输出。格式字符串中可以设置多个参数,例如格式字符串为"Failed to visit %s.","%s",参数类型为 string 的变参标识,具体取值在 args 中定义。每个参数需添加隐私标识,分为{public}或{private},默认为{private}。{public}表示日志打印结果可见;{private}表示日志打印结果不可见,输出结果为<private>。

3)参数 args:可以为 0 个或多个参数,是格式字符串中参数类型对应的参数列表。参数的数量、类型必须与格式字符串中的标识一一对应。

以输出一条 WARN 级别的信息为例,示例代码如下。

```
HiLog.warn(LABEL, "Failed to visit %{private}s, reason:%{public}d.", url, errno);
```

该行代码表示输出一个日志标签为 LABEL 的警告信息，格式字符串为："Failed to visit %{private}s, reason:%{public}d."。其中变参 url 的格式为私有的字符串，errno 为公共的整型数。

3．查看日志信息

DevEco Studio 提供了"Log > HiLog"窗口查看日志信息，开发者可通过设置设备、进程、日志级别和搜索关键词来筛选日志信息。搜索功能支持使用正则表达式，开发者可通过搜索自定义的业务领域值和标签来筛选日志信息。

如图 11-3 所示，根据实际情况选择了设备和进程后，搜索业务领域值"00201"进行筛选。

图 11-3　HiLog 窗口

得到对应的日志信息。

```
    07-19 14:22:56.071 13849-13849/com.example.myapplication W 00201/MY_TAG: Failed to visit , reason:503.
```

1) W 表示日志级别为 WARN。

2) 00201/MY_TAG 为开发者在 HiLogLabel 中定义的内容。

3) 日志内容中的 url 为私有参数，不显示具体内容，仅显示<private>。errno 为公有参数，显示实际取值 503。

HiLog 窗口左侧常用按钮的作用如图 11-4 所示。

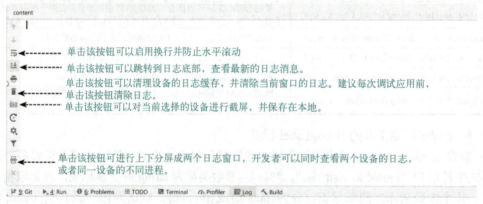

图 11-4　HiLog 窗口常用按钮

11.2.2　更新项目代码

将项目中的所有 System.out.println 方式打印的日志全部替换成 HiLog，使用有等级的方式进行打印，可以提高程序的性能。如图 11-5 所示，选择"Edit"→"Find"→"Find in Files"。

图 11-5　Find in Files

如图 11-6 所示，在 Find in Files 中搜索 System.out.println，然后使用 HiLog 逐步替换。

图 11-6　搜索 System.out.println

替换操作总共需要更新 3 个 java 文件：MainAbilitySlice.java、CloudParameterSettingAbility-Slice.java 和 LoginAbilitySlice.java。在每个文件类的属性中定义如下常量。

```
// 定义日志标签
private static final HiLogLabel LOG_LABEL = new HiLogLabel(HiLog.LOG_APP, 0, "类名");
```

其中"类名"使用当前的类名替换。

在需要注释的地方,使用

```
HiLog.debug(LOG_LABEL, 字符串);
```

其中"字符串"即为要打印的内容。

11.2.3 编译运行

编译运行程序,进行日志过滤,如图 11-7 所示,

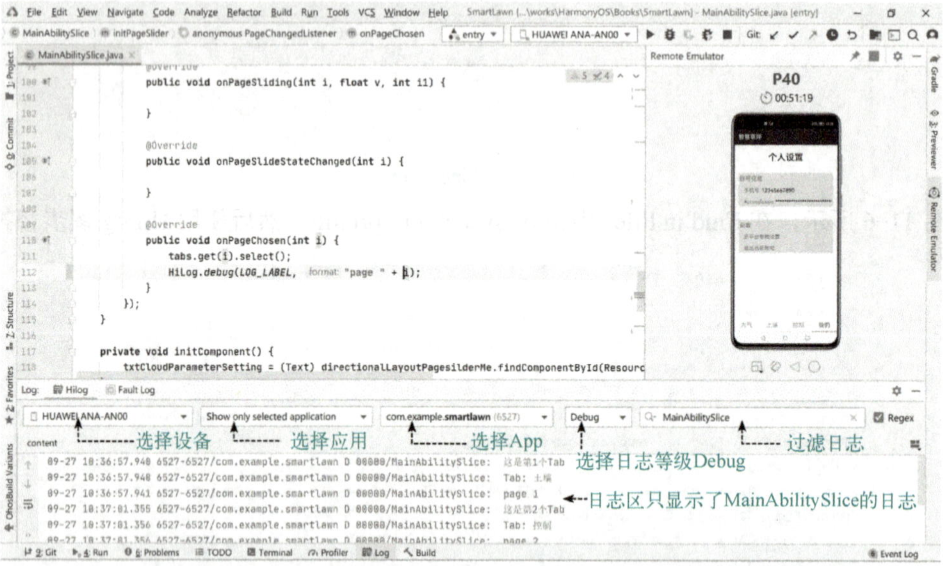

图 11-7　HiLog 过滤日志

11.2.4 提交代码到仓库

将此功能的代码提交到本地仓库,并添加日志"使用 HiLog 替换 System.out.println 打印日志"。

11.3　从云平台获取传感器数据

用户切换到大气界面或者土壤界面时,启动 3s 定时任务,即每隔 3s 从云平台获取传感器数据,然后在 UI 线程中更新界面数据。

11.3.1　更新 MyApplication.java

更新 MyApplication.java 文件,修改 readAccount 方法,增加 telephone 变量赋值,此处 telephone 和账号一致。readAccount 代码如下。

```
1.   public void readAccount() {
2.       isLogin = preferences.getBoolean("isLogin", false);
```

```
3.        if (isLogin) {
4.            account = preferences.getString("account", "");
5.            password = preferences.getString("Password", "");
6.            accesstoken = preferences.getString("AccessToken", "");
7.            telephone = account;
8.        }
9.    }
```

11.3.2　创建 SensorData.java

在 bean 目录下，创建 SensorData.java 文件。该文件内类的属性与新大陆物联网云平台的 RESTful API 模糊查询传感器数据响应属性一致。文件代码如下。

```
1.  {
2.      "ResultObj": [
3.          {
4.              "Unit": "lx",
5.              "ApiTag": "z_light",
6.              "Groups": 1,
7.              "Protocol": 2,
8.              "Name": "光照",
9.              "CreateDate": "2021-07-13 17:17:51",
10.             "TransType": 0,
11.             "DataType": 1,
12.             "TypeAttrs": "",
13.             "DeviceID": 309241,
14.             "SensorType": "light",
15.             "GroupID": null,
16.             "Coordinate": null,
17.             "Value": 10000.0,
18.             "RecordTime": "2021-09-09 08:59:50"
19.         },
20.         {
21.             "Unit": "℃",
22.             "ApiTag": "z_tpt",
23.             "Groups": 1,
24.             "Protocol": 2,
25.             "Name": "温度",
26.             "CreateDate": "2021-07-13 17:19:06",
27.             "TransType": 0,
28.             "DataType": 1,
29.             "TypeAttrs": "",
30.             "DeviceID": 309241,
31.             "SensorType": "temperature",
32.             "GroupID": null,
33.             "Coordinate": null,
34.             "Value": 15.59,
35.             "RecordTime": "2021-09-09 08:59:50"
```

```
36.         }
37.     ],
38.     "Status": 0,
39.     "StatusCode": 0,
40.     "Msg": null,
41.     "ErrorObj": null
42. }
```

SensorData.java 完整代码如下。

```
1.  package com.example.smartagriculture.bean;
2.
3.  import java.io.Serializable;
4.  import java.util.List;
5.
6.  public class SensorData implements Serializable {
7.      public List<ResultObj> ResultObj;
8.      public static class ResultObj implements Serializable{
9.          public String Unit;
10.         public String ApiTag;
11.         public int Groups;
12.         public int Protocol;
13.         public String Name;
14.         public String CreateDate;
15.         public int TransType;
16.         public int DataType;
17.         public String TypeAttrs;
18.         public int DeviceID;
19.         public String SensorType;
20.         public int GroupID;
21.         public String Coordinate;
22.         public String Value;
23.         public String RecordTime;
24.     }
25.     public int Status;
26.     public int StatusCode;
27.     public String Msg;
28.     public String ErrorObj;
29. }
```

11.3.3 更新 Wan.java

在 net 目录下更新 Wan.java 文件，增加 getSensorData，代码如下。

```
@GET("devices/{deviceId}/sensors")
Call<SensorData> getSensorData(@Header("AccessToken") String accessToken,
            @Path("deviceId") int deviceId, @QueryMap Map<String,
            String> sensorIDs);
```

11.3.4 更新 MainAbilitySlice.java

1. 修改成员属性

```
1.  private Text txtCloudParameterSetting, txtQuitAccount, txtTelephone-
Value, txtAccessTokenValue;
2.  private Timer timer;
3.  private TimerTask taskAir;
4.  //大气环境监控 UI 组件
5.  private Text txtTemperatureValue, txtHumidityValue, txtWindSpeedValue,
txtWindDirectionValue, txtIlluminationValue,
6.      txtAtmosValue, txtPM2_5Value, txtCarbonDioxideValue,
txtDataSyncDatetimeAirValue;
7.  //参数持久化
8.  private Preferences preferences;
9.  private MyApplication application;
10.
11. ToastDialog toastDialog;
12. SimpleDateFormat df;
```

第 2 行,增加定时器成员属性。

第 3 行,增加定时任务成员属性,用于定时从云平台获取传感器数据。

第 5、6 行,增加大气环境界面的数据组件。

第 8、9 行,增加持久化的成员属性。

第 12 行,增加 SimpleDateFormat 对象。

2. 修改 onStart 方法

```
1.  public void onStart(Intent intent) {
2.      super.onStart(intent);
3.      super.setUIContent(ResourceTable.Layout_ability_main);
4.      toastDialog = new ToastDialog(getContext()).setAlignment
(LayoutAlignment.CENTER);
5.      df = new SimpleDateFormat("yyyy-MM-dd HH:mm:ss");//设置日期格式
6.      application = (MyApplication)getAbility().getAbilityPackage();
7.      preferences = application.getPreferences();
8.      timer = new Timer();
9.      initTabList();
10.     initPageSlider();
11.     initComponent();
12.     initClickedListener();
13.     initTimerTask();
14.     int pagenum = pageSlider.getCurrentPage();
15.     if(0 == pagenum || 1 == pagenum){
16.         startTimer(pagenum);
17.     }
18. }
```

第 14~17 行,根据当前页面,启动对应的定时器,定时从云平台获取数据。

3. 修改 initPageSlider 方法中的 onPageChosen 方法

```
1.  @Override
2.  public void onPageChosen(int i) {
3.      tabs.get(i).select();
4.      HiLog.debug(LOG_LABEL, "选中page " + i);
5.      if(0 == i || 1 == i) {
6.          startTimer(i);
7.      }
8.      if(3 == i) {
9.          updateMePage();
10.     }
11. }
```

第 5~7 行，页面切换时，根据页面位置，启动对应定时器任务。

第 8~10 行，当页面切换到"我的"页面时，调用 **updateMePage** 方法。

4. 修改 initComponent 方法

```
1.  private void initComponent() {
2.      //我的组件获取
3.      txtCloudParameterSetting = (Text) directionalLayoutPagesliderMe.findComponentById(ResourceTable.Id_txtCloudParameterSetting);
4.      txtTelephoneValue = (Text)directionalLayoutPagesliderMe.findComponentById(ResourceTable.Id_txtTelephoneValue);
5.      txtAccessTokenValue = (Text)directionalLayoutPagesliderMe.findComponentById(ResourceTable.Id_txtAccessTokenValue);
6.      txtQuitAccount = (Text)directionalLayoutPagesliderMe.findComponentById(ResourceTable.Id_txtQuit);
7.      //大气环境监控 UI 组件获取
8.      txtTemperatureValue = (Text)dependentLayoutPagesliderAir.findComponentById(ResourceTable.Id_txtTemperatureValue);
9.      txtHumidityValue = (Text)dependentLayoutPagesliderAir.findComponentById(ResourceTable.Id_txtHumidityValue);
10.     txtWindSpeedValue = (Text)dependentLayoutPagesliderAir.findComponentById(ResourceTable.Id_txtWindSpeedValue);
11.     txtWindDirectionValue = (Text)dependentLayoutPagesliderAir.findComponentById(ResourceTable.Id_txtWindDirectionValue);
12.     txtIlluminationValue = (Text)dependentLayoutPagesliderAir.findComponentById(ResourceTable.Id_txtIlluminationValue);
13.     txtAtmosValue = (Text)dependentLayoutPagesliderAir.findComponentById(ResourceTable.Id_txtAtmosValue);
14.     txtPM2_5Value = (Text)dependentLayoutPagesliderAir.findComponentById(ResourceTable.Id_txtPM2_5Value);
15.     txtCarbonDioxideValue = (Text)dependentLayoutPagesliderAir.findComponentById(ResourceTable.Id_txtCarbonDioxideValue);
16.     txtDataSyncDatetimeAirValue = (Text)dependentLayoutPagesliderAir.findComponentById(ResourceTable.Id_txtDataSyncDatetimeAirValue);
17. }
```

获取"我的"页面与"大气监控"页面的组件。

5. 增加新的方法

```
1.  private void initTimerTask() {
2.      initTimerAirTask();
3.  }
4.  private void initTimerAirTask() {
5.      taskAir = new TimerTask() {
6.          @Override
7.          public void run() {
8.              Map<String, String> sensorIDs = new HashMap<String,String>();
9.              if(application.isLogin() && application.isCloudParameterSetting()){
10.                 HiLog.debug(LOG_LABEL, application.getWindDirectionID());
11.                 HiLog.debug(LOG_LABEL, application.getCarbonDioxideID());
12.                 sensorIDs.put("apiTags", application.getTemperatureID() + "," +
13.                         application.getHumidityID() + "," + application.getWindSpeedID() + "," +
14.                         application.getWindDirectionID() + "," + application.getIlluminationID() + "," +
15.                         application.getAtmosID() + "," + application.getPm2_5ID() + "," +
16.                         application.getCarbonDioxideID());
17.                 application.getmWan().getSensorData(application.getAccesstoken(),
18.                         Integer.parseInt(application.getDeviceId()), sensorIDs)
19.                         .enqueue(new Callback<SensorData>() {
20.                             @Override
21.                             public void onResponse(Call<SensorData> call, Response<SensorData> response) {
22.                                 updateAirData(response.body());
23.                             }
24.                             @Override
25.                             public void onFailure(Call<SensorData> call, Throwable throwable) {
26.                                 toastDialog.setText(throwable.getMessage())
27.                                         .show();
28.                             }
29.                         });
30.             }
31.         }
32.     };
33. }
34. private void startTimer(int pagenum) {
```

```
35.        timer.cancel();
36.        timer = new Timer();
37.        if(0 == pagenum){
38.            initTimerAirTask();
39.            timer.schedule(taskAir, 0, 3000);
40.        }
41. }
42. private void updateAirData(SensorData body) {
43.        String temperature = "", humidity = "", windSpeed = "", windDirection = "", illumination = "", atmos = "", pm2_5 = "", carbonDioxide = "", datetime;
44.        List<String> list = new ArrayList<>();
45.        for(SensorData.ResultObj data: body.ResultObj){
46.            String apiTag = data.ApiTag;
47.            if(apiTag.equals(application.getTemperatureID())){
48.                temperature = data.Value;
49.            }else if(apiTag.equals(application.getHumidityID())){
50.                humidity = data.Value;
51.            }else if(apiTag.equals(application.getWindSpeedID())){
52.                windSpeed = data.Value;
53.            }else if(apiTag.equals(application.getWindDirectionID())){
54.                windDirection = data.Value;
55.            }else if(apiTag.equals(application.getIlluminationID())){
56.                illumination = data.Value;
57.            }else if(apiTag.equals(application.getAtmosID())) {
58.                atmos = data.Value;
59.            }else if(apiTag.equals(application.getPm2_5ID())) {
60.                pm2_5 = data.Value;
61.            }else if(apiTag.equals(application.getCarbonDioxideID())){
62.                carbonDioxide = data.Value;
63.            }
64.        }
65.        datetime = df.format(new Date());// new Date()为获取当前系统时间
66.        list.add(temperature);
67.        list.add(humidity);
68.        list.add(windSpeed);
69.        list.add(windDirection);
70.        list.add(illumination);
71.        list.add(atmos);
72.        list.add(pm2_5);
73.        list.add(carbonDioxide);
74.        list.add(datetime);
75.        updateAirUI(list);
76. }
77. private void updateAirUI(List<String> list){
78.        getUITaskDispatcher().asyncDispatch(new Runnable() {
```

```
79.          @Override
80.          public void run() {
81.              txtTemperatureValue.setText(list.get(0));
82.              txtHumidityValue.setText(list.get(1));
83.              txtWindSpeedValue.setText(list.get(2));
84.              txtWindDirectionValue.setText(list.get(3));
85.              txtIlluminationValue.setText(list.get(4));
86.              txtAtmosValue.setText(list.get(5));
87.              txtPM2_5Value.setText(list.get(6));
88.              txtCarbonDioxideValue.setText(list.get(7));
89.              txtDataSyncDatetimeAirValue.setText(list.get(8));
90.          }
91.      });
92. }
93. private void updateMePage() {
94.      String telephone = application.getTelephone();
95.      getUITaskDispatcher().asyncDispatch(new Runnable() {
96.          @Override
97.          public void run() {
98.              txtTelephoneValue.setText(telephone);
99.          }
100.     });
101. }
```

第 1~3 行，initTimerTask 方法的实现，调用 initTimerAirTask 大气监控的任务。

第 4~33 行，initTimerAirTask 方法的实现。

第 5 行，创建定时任务。

第 8 行，创建 Map 对象，存储 apiTags 映射。

第 9 行，判断用户是否已经登录，云平台参数是否已经设置。

第 12 行，设置 Map 内容，存储 apiTags 映射，包含云平台传感器的 ID，字符串格式以 "," 间隔。

第 17 行，发起网络请求。

第 22 行，调用 updateAirData 方法，传递云平台响应内容。

第 34~41 行，startTimer 方法的实现。

第 35 行，删除定时器。

第 36 行，创建新的定时器。

第 37~40 行，如果当前页面是大气环境监控页面，则调用 initTimerAirTask 方法，初始化大气环境的定时任务，然后开始以 3s 为周期执行获取云平台传感器数据的任务。

第 42~76 行，updateAirData 方法的实现，从云平台响应体里获取传感器数据值。并获取当前日期时间，然后将数据加入列表中，调用 updateAirUI 方法。

第 77~92 行，updateAirUI 方法的实现。在 UI 线程里，更新界面的数据。

第 93~101 行，updateMePage 方法的实现，从 application 对象里获取"手机号"信息，并在 UI 线程里更新界面。

11.3.5 编译运行

连接远程模拟器，编译运行，结果如图 11-8 所示，发现风向数据未能获取。

11.3.6 调试解决 Bug

打开日志，查看 HiLog 日志窗口，如图 11-9 所示，应用请求了 8 个传感器数据，而云平台只响应了 7 个传感器数据，其中缺少了风向，这正好和我们的运行结果吻合，说明程序运行正常，问题可能出现在云平台。

继续检查云平台上的数据，如图 11-10 所示，可以发现问题所在，即风向传感器 ID 与 App 端不一致，可以采取改写更改任何一方的 ID 来修复，本文更改了 App 端风向传感器 ID，可以看到如图 11-11 所示，风向传感器数据立刻获取到。切换到"我的"页签，可看到"手机号"信息更新正常，如图 11-12 所示。

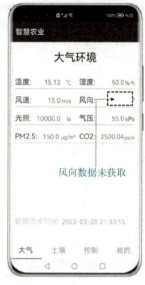

图 11-8 传感器数据获取

图 11-9 响应日志

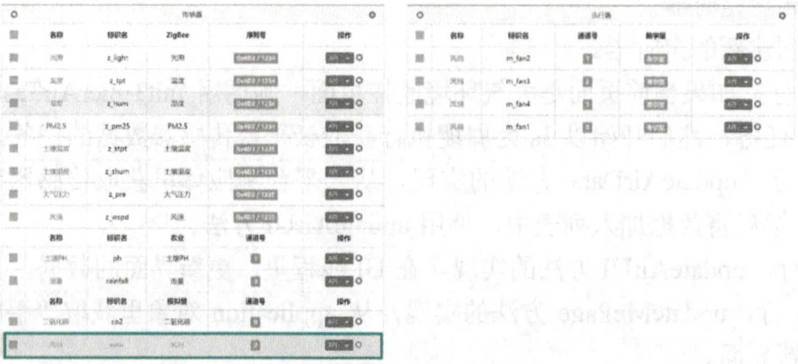

图 11-10 云平台传感器参数

图 11-11　风向数据获取

图 11-12　"我的"页签数据

11.4　提交代码到仓库

将任务 11 的代码提交到本地仓库，完整流程如下。

1）通过"git status"查看当前目录状态。

2）通过"git add."将当前目录下所有改动加入暂存区。

3）通过"git commit -m 从云平台获取传感器数据"，提交暂存区内容到本地版本库。

4）通过"git status"查看当前工作目录，可以看到目录干净。

5）通过"git tag -a task11 -m 从云平台获取传感器数据"，创建任务 11 标签。

6）通过"git log --pretty=oneline"，查看版本库日志，可以看到提交的最新内容，以及 task11 标签，即任务 11 的版本。

任务 12　创建土壤监控界面

任务概述

如何在土壤环境监控界面中显示传感器的实时数据？在任务 11 中已经实现了大气环境监控界面数据的获取。图 12-1 所示为土壤环境监控的界面，本任务主要完成土壤环境监控界面的设计以及逻辑的实现，实现传感器数据的获取并显示。

图 12-1　土壤监控界面

知识目标

- 掌握嵌套布局的设计。
- 掌握兼葭（JianJia）网络库的使用。
- 掌握 PageSlider 和 TabList 页签之间的切换。

能力目标

- 综合利用前面已学知识进行设计与实现。

12.1　更新 pageslider_soil.xml 界面

修改 pageslider_soil.xml，根据图 12-1，充分利用嵌套布局的方法设计土壤监控界面。完整代码可以参考本书配套代码，在代码目录通过以下命令提取任务 12 的代码版本。

```
git checkout task12
```

12.2　更新 MainAbilitySlice.java 代码

1. 修改成员属性

在原先的 taskAir 后面增加 taskSoil 声明。

```
         private TimerTask taskAir, taskSoil;
```
增加 MainAbilitySlice 成员属性土壤监控 UI 组件。

//土壤环境监控 UI 组件
```
         private Text txtPHValue, txtRainfallValue, txtSoilTemperatureValue,
             txtSoilHumidityValue, txtDataSyncDatetimeSoilValue;
```

2. 更新 initComponent 方法

在方法末尾增加与土壤监控相关的初始化。

```
 1.     //土壤环境监控 UI 组件获取
 2.     txtPHValue = (Text)dependentLayoutPagesliderSoil.findComponentById(ResourceTable.Id_txtPHValue);
 3.     txtRainfallValue = (Text)dependentLayoutPagesliderSoil.findComponentById(ResourceTable.Id_txtRainfallValue);
 4.     txtSoilTemperatureValue = (Text)dependentLayoutPagesliderSoil.findComponentById(ResourceTable.Id_txtSoilTemperatureValue);
 5.     txtSoilHumidityValue = (Text)dependentLayoutPagesliderSoil.findComponentById(ResourceTable.Id_txtSoilHumidityValue);
 6.     txtDataSyncDatetimeSoilValue = (Text)dependentLayoutPagesliderSoil.findComponentById(ResourceTable.Id_txtDataSyncDatetimeSoilValue);
```

3. 增加方法

创建 initTimerSoilTask、updateSoilData、updateSoilUI 方法，类比任务 11 中的方法。

```
 1.     private void initTimerSoilTask() {
 2.             taskSoil = new TimerTask() {
 3.                 @Override
 4.                 public void run() {
 5.                     Map<String, String> sensorIDs = new HashMap<String, String>();
 6.                     if(application.isLogin() && application.isCloudParameterSetting()){
 7.                         sensorIDs.put("apiTags", application.getPHID() + "," +
 8.                                 application.getRainfallID() + "," + application.getSoilTemperatureID() + "," +
 9.                                 application.getSoilHumidityID());
10.                         application.getmWan().getSensorData(application.getAccesstoken(), Integer.parseInt(application.getDeviceId()), sensorIDs)
11.                                 .enqueue(new Callback<SensorData>() {
12.                                     @Override
13.                                     public void onResponse(Call<SensorData> call, Response<SensorData> response) {
14.                                         updateSoilData(response.body());
15.                                     }
16.
17.                                     @Override
```

```
18.                              public void onFailure(Call<SensorData> call, Throwable throwable) {
19.                                  toastDialog.setText(throwable.getMessage())
20.                                          .show();
21.                              }
22.                          });
23.
24.                  }
25.              }
26.          };
27.      }
28.
29.      private void updateSoilData(SensorData body) {
30.          String ph = "", rainfall = "", soil_temperature = "", soil_humidity = "", datetime;
31.          List<String> list = new ArrayList<>();
32.          for(SensorData.ResultObj data: body.ResultObj){
33.              String apiTag = data.ApiTag;
34.              if(apiTag.equals(application.getPHID())){
35.                  ph = data.Value;
36.              }else if(apiTag.equals(application.getRainfallID())){
37.                  rainfall = data.Value;
38.              }else if(apiTag.equals(application.getSoilTemperatureID())){
39.                  soil_temperature = data.Value;
40.              }else if(apiTag.equals(application.getSoilHumidityID())){
41.                  soil_humidity = data.Value;
42.              }
43.          }
44.          SimpleDateFormat df = new SimpleDateFormat("yyyy-MM-dd HH:mm:ss");//设置日期格式
45.          datetime = df.format(new Date());// new Date()为获取当前系统时间
46.          toastDialog.setText("datetime" + datetime);
47.          list.add(ph);
48.          list.add(rainfall);
49.          list.add(soil_temperature);
50.          list.add(soil_humidity);
51.          list.add(datetime);
52.          updateSoilUI(list);
53.      }
54.
55.      private void updateSoilUI(List<String> list) {
56.          getUITaskDispatcher().asyncDispatch(new Runnable() {
57.              @Override
58.              public void run() {
59.                  txtPHValue.setText(list.get(0));
```

```
60.                    txtRainfallValue.setText(list.get(1));
61.                    txtSoilTemperatureValue.setText(list.get(2));
62.                    txtSoilHumidityValue.setText(list.get(3));
63.                    txtDataSyncDatetimeSoilValue.setText(list.get(4));
64.              }
65.          });
66.     }
```

4. 更新 initTimerTask 方法

```
1.   private void initTimerTask() {
2.        initTimerAirTask();
3.        initTimerSoilTask();
4.   }
```

在 initTimerAirTask 下面增加 initTimerSoilTask 方法的调用。

5. 更新 startTimer 方法

```
1.   private void startTimer(int pagenum) {
2.        timer.cancel();
3.        timer = new Timer();
4.        if(0 == pagenum){
5.            initTimerAirTask();
6.            timer.schedule(taskAir, 0, 3000);
7.        }else if(1 == pagenum) {
8.            initTimerSoilTask();
9.            timer.schedule(taskSoil, 0, 3000);
10.       }
11.  }
```

根据当前页面位置，启动对应的定时器周期性任务，定时从云平台获取传感器数据。

12.3 更新 string.json

在 base/element 目录下更新 string.json，增加字符串，如表 12-1 所示。

表 12-1　字符串表

name	value	name	value	name	value
strTxtPH	pH 值	strTxtRainfall	雨量	strTxtSoilTemperature	温度
strTxtSoilHumidity	湿度	strTxtSoilEnvironment	土壤环境		

12.4 编译运行

连接远程模拟器，编译运行，结果如图 12-2 所示。

图 12-2　土壤环境数据

12.5　提交代码到仓库

将任务 12 的代码提交到本地仓库，完整流程如下。

1）通过"git status"查看当前目录状态。
2）通过"git add."将当前目录下所有改动加入暂存区。
3）通过"git commit -m 创建土壤监控"，提交暂存区内容到本地版本库。
4）通过"git status"，查看当前工作目录，可以看到目录干净。
5）通过"git tag -a task12 -m 创建土壤监控"，创建任务 12 标签。
6）通过"git log --pretty=oneline"查看版本库日志，可以看到提交的最新内容，以及 task12 标签，即任务 12 的版本。

任务 13　创建执行器控制

任务概述

执行器在硬件连接上与传感器方式不同，但对于 App 来说是透明的，App 只需和物联网云平台进行通信即可控制执行器。本任务主要完成执行器界面的设计与逻辑的实现，如图 13-1 所示。

图 13-1　控制界面

知识目标

- 掌握 Switch 组件的使用。
- 掌握云平台命令发送的 RESTful API。
- 掌握蒹葭（JianJia）网络库的使用。

能力目标

- 能使用 Switch 组件进行设计。
- 能使用蒹葭（JianJia）发送命令。

13.1　使用 Switch 组件

Switch 是切换单个设置开/关两种状态的组件。Switch 的公有 XML 属性继承自 Text。其私有 XML 属性如表 13-1 所示。

表 13-1　Switch 的自有 XML 属性

属性名称	中文描述	取值	取值说明	使用案例
text_state_on	开启时显示的文本	string 类型	可以直接设置文本字串，也可以引用 string 资源（推荐使用）	ohos:text_state_on="联系" ohos:text_state_on="$string:test_str"
text_state_off	关闭时显示的文本	string 类型	可以直接设置文本字串，也可以引用 string 资源（推荐使用）	ohos:text_state_off="联系" ohos:text_state_off="$string:test_str"
track_element	轨迹样式	Element 类型	可直接配置色值，也可引用 color 资源或引用 media 或 graphic 下的图片资源	ohos:track_element="#FF0000FF" ohos:track_element="$color:black" ohos:track_element="$media:media_src" ohos:track_element="$graphic:graphic_src"

(续)

属性名称	中文描述	取值	取值说明	使用案例
thumb_element	thumb 样式	Element 类型	可直接配置色值，也可引用 color 资源或引用 media 或 graphic 下的图片资源	ohos:thumb_element="#FF0000FF" ohos:thumb_element="$color:black" ohos:thumb_element="$media:media_src" ohos:thumb_element="$graphic:graphic_src"
marked	当前状态（选中或未选中）	boolean 类型	可以直接设置 true 或 false，也可以引用 boolean 资源。true 则当前状态为选中，false 则当前状态为未选中	ohos:marked="true" ohos:marked="$boolean:true"
check_element	状态标志样式	Element 类型	可直接配置色值，也可引用 color 资源或引用 media 或 graphic 下的图片资源	ohos:check_element="#000000" ohos:check_element="$color:black" ohos:check_element="$media:media_src" ohos:check_element="$graphic:graphic_src"

13.2 更新 pageslider_control.xml 文件

1. 布局设计

更新 pageslider_control.xml 文件，代码如下。

```
1.  <?xml version="1.0" encoding="utf-8"?>
2.  <DirectionalLayout
3.      xmlns:ohos="http://schemas.huawei.com/res/ohos"
4.      ohos:height="match_parent"
5.      ohos:width="match_parent"
6.      ohos:background_element="$graphic:background_ability_main"
7.      ohos:orientation="vertical">
8.  
9.      <Text
10.         ohos:height="match_content"
11.         ohos:width="match_parent"
12.         ohos:bottom_margin="20vp"
13.         ohos:layout_alignment="horizontal_center"
14.         ohos:text="$string:strTxtControl"
15.         ohos:text_alignment="center"
16.         ohos:text_size="30vp"
17.         ohos:top_margin="20vp"/>
18.  
19.     <DirectionalLayout
20.         ohos:height="match_content"
21.         ohos:width="match_parent"
22.         ohos:bottom_margin="15vp"
23.         ohos:orientation="horizontal"
24.         ohos:top_margin="5vp">
25.  
26.         <DependentLayout
27.             ohos:height="match_content"
28.             ohos:width="0"
```

```
29.                ohos:background_element="$graphic:background_btn_switch"
30.                ohos:left_margin="15vp"
31.                ohos:right_margin="5vp"
32.                ohos:weight="1">
33.
34.                <Text
35.                    ohos:id="$+id:txtWaterValve1"
36.                    ohos:height="80vp"
37.                    ohos:width="match_content"
38.                    ohos:horizontal_center="true"
39.                    ohos:text="$string:strTxtWaterValve1"
40.                    ohos:text_color="#444647"
41.                    ohos:text_size="20fp"/>
42.
43.                <Switch
44.                    ohos:id="$+id:switchWaterValve1"
45.                    ohos:height="50vp"
46.                    ohos:width="80vp"
47.                    ohos:below="$id:txtWaterValve1"
48.                    ohos:bottom_padding="10vp"
49.                    ohos:horizontal_center="true"
50.                    ohos:text_color="#FFFFFF"
51.                    ohos:text_state_off="OFF"
52.                    ohos:text_state_on="ON"
53.                    ohos:text_size="16fp"
54.                    ohos:text_color_on="#FFFFFF"/>
55.            </DependentLayout>
56.
57.            <DependentLayout
58.                ohos:height="match_content"
59.                ohos:width="0"
60.                ohos:background_element="$graphic:background_btn_switch"
61.                ohos:left_margin="5vp"
62.                ohos:right_margin="15vp"
63.                ohos:weight="1">
64.
65.                <Text
66.                    ohos:id="$+id:txtWaterValve2"
67.                    ohos:height="80vp"
68.                    ohos:width="match_content"
69.                    ohos:horizontal_center="true"
70.                    ohos:text="$string:strTxtWaterValve2"
71.                    ohos:text_color="#444647"
72.                    ohos:text_size="20fp"/>
73.
74.                <Switch
75.                    ohos:id="$+id:switchWaterValve2"
```

```
76.            ohos:height="50vp"
77.            ohos:width="80vp"
78.            ohos:below="$id:txtWaterValve2"
79.            ohos:bottom_padding="10vp"
80.            ohos:horizontal_center="true"
81.            ohos:text_color="#FFFFFF"
82.            ohos:text_state_off="OFF"
83.            ohos:text_state_on="ON"
84.            ohos:text_size="16fp"
85.            ohos:text_color_on="#FFFFFF"/>
86.        </DependentLayout>
87.
88.    </DirectionalLayout>
89.
90.    <DirectionalLayout
91.        ohos:height="match_content"
92.        ohos:width="match_parent"
93.        ohos:bottom_margin="15vp"
94.        ohos:orientation="horizontal"
95.        ohos:top_margin="5vp">
96.
97.        <DependentLayout
98.            ohos:height="match_content"
99.            ohos:width="0"
100.           ohos:background_element="$graphic:background_btn_switch"
101.           ohos:left_margin="15vp"
102.           ohos:right_margin="5vp"
103.           ohos:weight="1">
104.
105.           <Text
106.               ohos:id="$+id:txtWaterValve3"
107.               ohos:height="80vp"
108.               ohos:width="match_content"
109.               ohos:horizontal_center="true"
110.               ohos:text="$string:strTxtWaterValve3"
111.               ohos:text_color="#444647"
112.               ohos:text_size="20fp"/>
113.
114.           <Switch
115.               ohos:id="$+id:switchWaterValve3"
116.               ohos:height="50vp"
117.               ohos:width="80vp"
118.               ohos:below="$id:txtWaterValve3"
119.               ohos:bottom_padding="10vp"
120.               ohos:horizontal_center="true"
121.               ohos:text_color="#FFFFFF"
122.               ohos:text_state_off="OFF"
```

```
123.                        ohos:text_state_on="ON"
124.                        ohos:text_size="16fp"
125.                        ohos:text_color_on="#FFFFFF"/>
126.            </DependentLayout>
127.
128.            <DependentLayout
129.                ohos:height="match_content"
130.                ohos:width="0"
131.                ohos:background_element="$graphic:background_btn_switch"
132.                ohos:left_margin="5vp"
133.                ohos:right_margin="15vp"
134.                ohos:weight="1">
135.
136.                <Text
137.                    ohos:id="$+id:txtWaterValve4"
138.                    ohos:height="80vp"
139.                    ohos:width="match_content"
140.                    ohos:horizontal_center="true"
141.                    ohos:text="$string:strTxtWaterValve4"
142.                    ohos:text_color="#444647"
143.                    ohos:text_size="20fp"/>
144.
145.                <Switch
146.                    ohos:id="$+id:switchWaterValve4"
147.                    ohos:height="50vp"
148.                    ohos:width="80vp"
149.                    ohos:below="$id:txtWaterValve4"
150.                    ohos:bottom_padding="10vp"
151.                    ohos:horizontal_center="true"
152.                    ohos:text_color="#FFFFFF"
153.                    ohos:text_state_off="OFF"
154.                    ohos:text_state_on="ON"
155.                    ohos:text_size="16fp"
156.                    ohos:text_color_on="#FFFFFF"/>
157.            </DependentLayout>
158.
159.        </DirectionalLayout>
160.
161.    </DirectionalLayout>
```

根据图 13-1，充分利用嵌套布局技术，设计控制界面的布局。其中，Switch 是开关组件，具有滑动效果。

2. 背景设计

在 base/graphic 目录下，创建 background_btn_switch.xml 背景设计文件，代码如下。

```
1.  <?xml version="1.0" encoding="UTF-8" ?>
2.  <shape xmlns:ohos="http://schemas.huawei.com/res/ohos"
3.         ohos:shape="rectangle">
```

```
4.     <corners ohos:radius="20"/>
5.     <stroke
6.         ohos:color="#FF699CE9"/>
7.     <solid
8.         ohos:color="#FF88E7E4"/>
9. </shape>
```

13.3 更新 java 文件

当用户单击开关时，App 发起网络请求，将命令发送给云平台，云平台与之相关的开关联动，同时向物联网感知层发送控制设备的命令。如果打开设备失败，则 App 端的开关自动切换到关闭状态。

13.3.1 创建 CmdRsp.java bean 文件

在 bean 目录下创建 CmdRsp.java 文件，该内容应与新大陆物联网云平台 RESTful API 中发送命令响应的格式匹配，如下。

```
1. {
2.     "ResultObj": "b3640dc9-aa60-413a-8112-8697a6125820",
3.     "Status": 0,
4.     "StatusCode": 0,
5.     "Msg": null,
6.     "ErrorObj": null
7. }
```

对应的 CmdRsp.java 代码如下。

```
1. package com.example.smartlawn.bean;
2.
3. import java.io.Serializable;
4.
5. public class CmdRsp implements Serializable {
6.     public String ResultObj;
7.     public int Status;
8.     public int StatusCode;
9.     public String Msg;
10.    public String ErrorObj;
11. }
```

13.3.2 更新 Wan.java

在 net 目录下更新 Wan.java，增加如下方法。

```
@POST("cmds")
Call<CmdRsp> sendCmd(@Header("AccessToken") String accessToken,
```

```
@QueryMap Map<String, String> urlData, @Body Object body);
```

13.3.3 更新 MainAbilitySlice.Java

1. 增加成员属性

```
//执行器 UI 组件
private Switch switchWaterValve1, switchWaterValve2, switchWaterValve3, switchWaterValve4;
private boolean bWaterValve1 = true, bWaterValve2 = true, bWaterValve3 = true, bWaterValve4 = true;
```

2. 修改 onStart 方法

在 initClickedListener 方法下面，增加 initSwitches 方法的调用。

```
1.    initClickedListener();
2.    initSwitches();
```

3. 更新 initComponent 方法

更新 initComponent 方法，在末尾增加执行器 UI 组件的获取。

```
1.    //执行器 UI 组件获取
2.    switchWaterValve1 = (Switch)directionalLayoutPagesilderControl.findComponentById(ResourceTable.Id_switchWaterValve1);
3.    switchWaterValve2 = (Switch)directionalLayoutPagesilderControl.findComponentById(ResourceTable.Id_switchWaterValve2);
4.    switchWaterValve3 = (Switch)directionalLayoutPagesilderControl.findComponentById(ResourceTable.Id_switchWaterValve3);
5.    switchWaterValve4 = (Switch)directionalLayoutPagesilderControl.findComponentById(ResourceTable.Id_switchWaterValve4);
```

4. 增加方法

分别创建 initSwitches、sendCmd、switchBack 三个方法，如下。

```
1.    private void initSwitches(){
2.        switchWaterValve1.setCheckedStateChangedListener(new AbsButton.CheckedStateChangedListener() {
3.            @Override
4.            public void onCheckedChanged(AbsButton absButton, boolean b) {
5.                if(!bWaterValve1){
6.                    bWaterValve1 = true;
7.                    return;
8.                }
9.                String waterValveId = application.getWaterValve1ID();
10.               if(b){
11.                   sendCmd("waterValve1", waterValveId, 1);
12.               }else {
13.                   sendCmd("waterValve1", waterValveId, 0);
```

```
14.            }
15.        }
16.    });
17.    switchWaterValve2.setCheckedStateChangedListener(new AbsButton.CheckedStateChangedListener() {
18.        @Override
19.        public void onCheckedChanged(AbsButton absButton, boolean b) {
20.            if(!bWaterValve2){
21.                bWaterValve2 = true;
22.                return;
23.            }
24.            String waterValveId = application.getWaterValve2ID();
25.            if(b){
26.                sendCmd("waterValve2", waterValveId, 1);
27.            }else {
28.                sendCmd("waterValve2", waterValveId, 0);
29.            }
30.        }
31.    });
32.    switchWaterValve3.setCheckedStateChangedListener(new AbsButton.CheckedStateChangedListener() {
33.        @Override
34.        public void onCheckedChanged(AbsButton absButton, boolean b) {
35.            if(!bWaterValve3){
36.                bWaterValve3 = true;
37.                return;
38.            }
39.            String waterValveId = application.getWaterValve3ID();
40.            if(b){
41.                sendCmd("waterValve3", waterValveId, 1);
42.            }else {
43.                sendCmd("waterValve3", waterValveId, 0);
44.            }
45.        }
46.    });
47.    switchWaterValve4.setCheckedStateChangedListener(new AbsButton.CheckedStateChangedListener() {
48.        @Override
49.        public void onCheckedChanged(AbsButton absButton, boolean b) {
50.            if(!bWaterValve4){
51.                bWaterValve4 = true;
52.                return;
53.            }
54.            String waterValveId = application.getWaterValve4ID();
55.            if(b){
56.                sendCmd("waterValve4", waterValveId, 1);
57.            }else {
```

```
58.                    sendCmd("waterValve4", waterValveId, 0);
59.                }
60.            }
61.        });
62.    }
63.
64.    private void sendCmd(String waterValve, String waterValveId, int cmd) {
65.        Map<String, String>  target = new HashMap<>();
66.        target.put("deviceId", application.getDeviceId());
67.        target.put("apiTag", waterValveId);
68.        boolean success = false;
69.        String finalWaterValve = waterValve;
70.        application.getmWan().sendCmd(application.getAccesstoken(), target, cmd).enqueue(new Callback<CmdRsp>() {
71.            @Override
72.            public void onResponse(Call<CmdRsp> call, Response<CmdRsp> response) {
73.                try {
74.                    if(response.body().Status == 0) {
75.                        String msg = "成功" + ((cmd == 1)?"打开":"关闭") + "水阀!";
76.                        toastDialog.setText(msg)
77.                                .show();
78.                    }else {
79.                        switchBack(finalWaterValve);
80.                        toastDialog.setText("出错:" + response.body().Msg)
81.                                .show();
82.                    }
83.                }catch (Exception e){
84.                    e.printStackTrace();
85.                    switchBack(finalWaterValve);
86.                    toastDialog.setText("访问出错!")
87.                            .show();
88.                }
89.            }
90.            @Override
91.            public void onFailure(Call<CmdRsp> call, Throwable throwable) {
92.                switchBack(finalWaterValve);
93.                toastDialog.setText("出错: " + throwable.getMessage())
94.                        .show();
95.            }
96.        });
97.    }
98.
99.    private void switchBack(String waterValve){
100.        switch (waterValve){
101.            case "waterValve1":
```

```
102.                bWaterValve1 = false;
103.                if (switchWaterValve1.isChecked()) {
104.                    switchWaterValve1.setChecked(false);
105.                } else {
106.                    switchWaterValve1.setChecked(true);
107.                }
108.                break;
109.            case "waterValve2":
110.                bWaterValve2 = false;
111.                if (switchWaterValve2.isChecked()) {
112.                    switchWaterValve2.setChecked(false);
113.                } else {
114.                    switchWaterValve2.setChecked(true);
115.                }
116.                break;
117.            case "waterValve3":
118.                bWaterValve3 = false;
119.                if (switchWaterValve3.isChecked()) {
120.                    switchWaterValve3.setChecked(false);
121.                } else {
122.                    switchWaterValve3.setChecked(true);
123.                }
124.                break;
125.            case "waterValve4":
126.                bWaterValve4 = false;
127.                if (switchWaterValve4.isChecked()) {
128.                    switchWaterValve4.setChecked(false);
129.                } else {
130.                    switchWaterValve4.setChecked(true);
131.                }
132.                break;
133.        }
134.    }
```

第 1~62 行，initSwitches 方法的实现，分别设置 4 个水阀开关的状态监听器。当开关状态变化时，进入 onCheckedChanged 处理流程。其中第 5 行，bWaterValve1 布尔值，初始化时为 true，代表开关命令的发送状态。当开关动作发生，如果与云平台交互异常，则通过该布尔值，避免再次向云平台发送命令，造成死循环。

第 64~97 行，sendCmd 方法的实现。

第 65~67 行，设置设备 ID 和执行器 ID。

第 70 行，发起网络请求。

第 72 行，响应成功处理流程。

第 74 行，如果 response.body().Status 为 0，代表命令发送成功，则弹出对话框提醒用户命令发送成功。

第 78 行，如果响应状态为非 0，则表示失败，进入失败处理流程。

第 79 行，调用 switchBack 方法，将开关状态反转，即恢复上一状态。

第 83~88 行，出现异常时，也调用 switchBack 方法，将开关状态反转，即恢复上一状态。

第 90~95 行，网络请求失败后，也调用 switchBack 方法，将开关状态反转，即恢复上一状态。

第 99~134 行，switchBack 方法的实现。其中第 102 行，将 bWaterValve1 赋值为 false，即开关状态恢复上一状态时，在状态变化处理流程里，不用向云平台发送命令，避免造成死循环。第 103~107 行，反转开关状态。

13.4 更新 string.json 文件

在 base/element 目录下更新 string.json 文件，增加字符串，如表 13-2 所示。

表 13-2 字符串表

name	value	name	value	name	value
strTxtWaterValve1	水阀 1	strTxtWaterValve2	水阀 2	strTxtWaterValve3	水阀 3
strTxtWaterValve4	水阀 4	strTxtControl	控制执行器		

13.5 编译运行

连接远程模拟器，编译运行，在云平台端使用风扇模拟水阀的动作。

13.5.1 打开水阀

如图 13-2 所示，App 端打开水阀，云平台端水阀（风扇）状态也打开，如图 13-3 所示，两者保持状态一致。

图 13-2 App 打开水阀

图 13-3 云平台水阀（风扇）打开

13.5.2 关闭水阀

如图 13-4 所示，App 端关闭水阀，云平台水阀（风扇）状态也关闭，如图 13-5 所示，两者保持状态一致。

图 13-4　App 关闭水阀

图 13-5　云平台水阀（风扇）关闭

13.5.3 设备未上线

如果云平台端设备未上线，则 App 端如图 13-6 所示，打开水阀时，提示设备未在线，同时 Switch 状态恢复关闭状态。

13.6　提交代码到仓库

图 13-6　设备未在线

将任务 13 的代码提交到本地仓库，完整流程如下。
1）通过 "git status" 查看当前目录状态。
2）通过 "git add ." 将当前目录下所有改动加入暂存区。
3）通过 "git commit -m 创建执行器控制"，提交暂存区内容到本地版本库。
4）通过 "git status" 查看当前工作目录，可以看到目录干净。
5）通过 "git tag -a task13 -m 创建执行器控制"，创建任务 13 标签。
6）通过 "git log --pretty=oneline" 查看版本库日志，可以看到提交的最新内容，以及 task13 标签，即任务 13 的版本。

任务 14　创建多语言环境

任务概述

本任务要完成 App 国际化时必须具备的功能——多语言支持，本 App 支持中、英文，跟随系统自动切换。当系统是英文环境时，App 显示英文；当系统是中文环境时，显示中文；当系统是其他环境时，显示中文。同时提供退出账号功能，当用户在"我的"界面，单击退出当前账号时，App 会自动清空配置文件和账号信息，跳转到登录界面。

知识目标

- 掌握多语言设置。
- 掌握 Operation 的 action 的用法。
- 掌握轻量级数据库的删除操作。
- 了解 Page Ability 生命周期。

能力目标

- 能使用多语言功能设计 App。
- 能删除轻量级数据存储的数据文件。
- 掌握 Page Ability 生命周期基本管理。

14.1　多语言设计

HarmonyOS 支持通过限定词目录实现多语言设计，需要使用以下三个文件。

（1）entry/src/main/resources/base/element/string.json

默认字符串引用文件，支持默认语言。

（2）entry/src/main/resources/en/element/string.json

英文环境下加载这个文件，支持英文语言。

（3）entry/src/main/resources/zh/element/string.json

中文环境下加载这个文件，支持中文语言。

1. 更新中文字符串文件

直接将默认语言文件内容复制到中文环境下的字符串文件里，可以实现系统在中文环境下使用中文显示。

2. 更新英语字符串文件

更新 en/element/string.json 文件，即将 zh/element/string.json 里的内容复制到 en/element/string.json，然后将中文翻译成英文，如表 14-1 所示。

表 14-1　字符串英文表

name	value	name	value	name	value
entry_MainAbility	entry_MainAbility	mainability_description	Java_Empty Ability	mainability_HelloWorld	Hello World
app_name	Smart Agriculture	splashability_description	Java_Empty Ability	splashability_HelloWorld	Hello World
entry_SplashAbility	entry_SplashAbility	strCancel	Cancel	strSmartAgriculture	Smart Agriculture

（续）

name	value	name	value	name	value
strTfTelephone	Please input telephone.	strTfPassword	Please input password.	strBtnLogin	Sign in->
strPgSoil	Soil	strPgAi	ATM	strPgControl	Control
strPgMe	Me	strTxtAirEnvironment	ATM Environment	strTxtTemperature	TEMP:
strTxtHumidity	HUM:	strTxtWindSpeed	WVEL:	strTxtWindDirection	WDir:
strTxtIllumination	ILL:	strTxtAtmos	Atmos:	strTxtPM2_5	PM2.5:
strTxtCarbonDioxide	CO_2:	strTxtDataSyncDatetime	Sync Time:	strTxtAccount	Account Information
strTxtTelephone	Telephone:	strTxtAccessToken	AccessToken:	strTxtSetting	Setting
strTxtCloudParametersSetting	Cloud Parameters Setting	strTxtQuit	Quit Account	strTxtMe	Profile
strTxtDeviceId	DeviceID:	strTxtAirSetting	Atmosphere Environment Setting	strTxtTemperatureId	TEMPID:
strTxtHumidityId	HUMID:	strTxtWindSpeedId	WVELID:	strTxtWindDirectionId	WDirID:
strTxtIlluminationId	ILLID:	strTxtAtmosId	AtmosID:	strTxtPM2_5Id	PM2.5ID:
strTxtCarbonDioxideId	CO_2ID:	strTxtSoilSetting	Soil Environment Setting	strTxtPHId	PHID:
strTxtRainfallId	RainfallID:	strTxtSoilTemperatureId	TEMPID:	strTxtSoilHumidityId	HumID:
strTxtControlSetting	Control Sensors Setting	strTxtWaterValve1Id	WV1ID:	strTxtWaterValve2Id	WV2ID:
strTxtWaterValve3Id	WV3ID:	strTxtWaterValve4Id	WV4ID:	strBtnSetting	Save
strTxtSoliEnvironment	Soil Environment	strTxtPH	pH:	strTxtRainfall	Rainfall:
strTxtSoilTemperature	Temp:	strTxtSoilHumidity	Hum:	strTxtControl	Control
strTxtWaterValve1	WaterValve1	strTxtWaterValve2	WaterValve2	strTxtWaterValve3	WaterValve3
strTxtWaterValve4	WaterValve4				

3．编译运行

编译运行程序，将系统语言设置为英语，程序效果如图 14-1 和图 14-2 所示。

图 14-1　欢迎界面

图 14-2　参数设置界面

4. 提交代码到仓库

将此功能的代码提交到本地仓库，并添加日志"创建多语言"。

14.2 全屏显示

App 界面一般不需要标题栏，可以通过设置让 App 全屏显示。

1. 修改配置文件

通过更改主题可以实现去除标题栏，增加下面代码。

```
"metaData": {
  "customizeData": [
    {
      "name": "hwc-theme",
      "extra": "",
      "value": "androidhwext:style/Theme.Emui.Light.NoTitleBar"
    }
  ]
}
```

将上面代码配置到 config.json 中，加入到 abilities JSON 数组的每个元素中。

2. 编译运行

编译运行程序，如图 14-3 所示，程序全屏显示，没有标题。

图 14-3　全屏显示

3. 提交代码到仓库

将此功能的代码提交到本地仓库，并添加日志"全屏显示"。

14.3 退出当前账号

App 如果提供账号登录功能，就应该提供退出账号的功能，以便用户切换账号。

14.3.1 更新 MyApplication.java 文件

1. 修改成员属性

增加成员属性。

```
DatabaseHelper databaseHelper;
```

初始化如下成员属性。

```
    private String account = "", password = "", telephone = "", college = "", role = "", accesstoken = "";
    private boolean isLogin = false, isCloudParameterSetting = false;
    private String deviceId = "";
    //大气环境传感器 ID 参数
    private String temperatureID = "", humidityID = "", windSpeedID = "", windDirectionID = "", illuminationID = "", atmosID = "", pm2_5ID = "", carbonDioxideID = "";
    //土壤环境传感器 ID 参数
    private String PHID = "", rainfallID = "", soilTemperatureID = "", soilHumidityID = "";
    //执行器传感器 ID
    private String waterValve1ID = "", waterValve2ID = "", waterValve3ID = "", waterValve4ID = "";
```

2. 修改 onInitialize 方法

在 onInitialize 中对 databaseHelper 进行初始化。

```
databaseHelper = new DatabaseHelper(this);
```

3. 增加方法

增加 getDatabaseHelper、initParams 和 getPREFERENCES_FILE_CLOUD_PARAMETERS 方法。

```
    public String getPREFERENCES_FILE_CLOUD_PARAMETERS() {
        return PREFERENCES_FILE_CLOUD_PARAMETERS;
    }

    public void initParams() {
        account = ""; password = ""; telephone = ""; college = ""; role = ""; accesstoken = "";
        isLogin = false; isCloudParameterSetting = false;
        deviceId = "";
        //大气环境传感器 ID 参数
        temperatureID = ""; humidityID = ""; windSpeedID = ""; windDirection-
```

```
ID = ""; illuminationID = "";
        atmosID = ""; pm2_5ID = ""; carbonDioxideID = "";
        //土壤环境传感器 ID 参数
        PHID = ""; rainfallID = ""; soilTemperatureID = ""; soilHumidityID = "";
        //执行器传感器 ID
        waterValve1ID = ""; waterValve2ID = ""; waterValve3ID = ""; waterVal-
ve4ID = "";
        preferences = databaseHelper.getPreferences(PREFERENCES_FILE_CLOUD_
PARAMETERS);
    }

    public DatabaseHelper getDatabaseHelper() {
        return databaseHelper;
    }
```

14.3.2 更新 SplashAbilitySlice.java 文件

修改 goToLogin 方法，增加第 12 行，跳转到 MainAbility 时，就可以结束当前的 Ability。

```
1.    private void goToLogin() {
2.        if (application.isLogin()) {
3.            Intent intent = new Intent();
4.            // 指定待启动 FA 的 bundleName 和 abilityName
5.            Operation operation = new Intent.OperationBuilder()
6.                    .withDeviceId("")
7.                    .withBundleName("com.example.smartagriculture")
8.                    .withAbilityName("com.example.smartagriculture.MainAbility")
9.                    .build();
10.           intent.setOperation(operation);
11.           startAbility(intent);
12.           terminateAbility();
13.       } else {
14.           Intent intent = new Intent();
15.           present(new LoginAbilitySlice(), intent);
16.       }
17.   }
```

14.3.3 更新 LoginAbilitySlice.java 文件

修改 goToMainAbility 方法，增加第 12 行，跳往 MainAbility 时，即可结束当前 Ability。

```
1.    public void goToMainAbility() {
2.        HiLog.debug(LOG_LABEL, "前往主界面");
3.        Intent intent = new Intent();
4.        // 指定待启动 FA 的 bundleName 和 abilityName
5.        Operation operation = new Intent.OperationBuilder()
6.                .withDeviceId("")
7.                .withBundleName("com.example.smartagriculture")
8.                .withAbilityName("com.example.smartagriculture.MainAbility")
9.                .build();
```

```
10.         intent.setOperation(operation);
11.         startAbility(intent);
12.         terminateAbility();
13.    }
```

14.3.4　更新 SplashAbility.java 文件

修改 SplashAbility.java，增加 Slice 路由的注册，在 onStart 方法末尾，增加如下代码。

```
        addActionRoute("action.login", LoginAbilitySlice.class.getName());
```

注册 LoginAbilitySlice 的路由，使得其他 Page Ability 可以导航到此 Slice。其中 action 为路由参数，位于 config.json 文件中。在 config.json 文件的 abilities 属性下，属于 com.example.smartagriculture.SplashAbility 元素的 skills 中的 actions 里添加如下第 8 行代码。

```
 1.   "skills": [
 2.     {
 3.       "entities": [
 4.         "entity.system.home"
 5.       ],
 6.       "actions": [
 7.         "action.system.home",
 8.         "action.login"
 9.       ]
10.     }
11.   ]
```

14.3.5　更新 MainAbilitySlice.java 文件

1. 修改 initClickedListener 方法

修改 initClickedListener 方法，增加 txtQuitAccount 组件监听单击事件，代码如下。

```
1.    private void initClickedListener() {
2.        txtCloudParameterSetting.setClickedListener(this);
3.        txtQuitAccount.setClickedListener(this);
4.    }
```

2. 修改 onClick 方法

修改 onClick 方法，增加 ResourceTable.Id_txtQuit 单击事件处理，增加 switch 新分支，代码如下。

```
1.    case ResourceTable.Id_txtQuit:{
2.        DatabaseHelper databaseHelper = application.getDatabaseHelper();
3.        boolean result = databaseHelper.deletePreferences(application.getPREFERENCES_FILE_CLOUD_PARAMETERS());
4.        application.initParams();
5.        Intent intent = new Intent();
6.        Operation operation = new Intent.OperationBuilder()
7.            .withAction("action.login")
```

```
8.                       .build();
9.              intent.setOperation(operation);
10.             startAbility(intent);
11.             terminateAbility();
12.             break;
13. }
```

第 3 行，删除轻量级数据存储的对象和配置文件。

第 4 行，重新初始化 MyApplication 子类的对象的属性。

第 5~10 行，导航到登录页。

第 11 行，关闭当前的 Ability。

terminateAbility 和 terminate 都属于手动管理 Page 与 AbilitySlice 生命周期的方法，下文将学习 Page 生命周期与 Page AbilitySlice 生命周期。

14.3.6　了解 Page Ability 生命周期

系统管理或用户操作等行为均会引起 Page 实例在其生命周期的不同状态之间的转换。Ability 类提供的回调机制能够让 Page 及时感知外界变化，从而正确地应对状态变化（比如释放资源），这有助于提升应用的性能和稳健性。

1．Page 生命周期回调

Page 生命周期的不同状态间的转换及其对应的回调如图 14-4 所示。

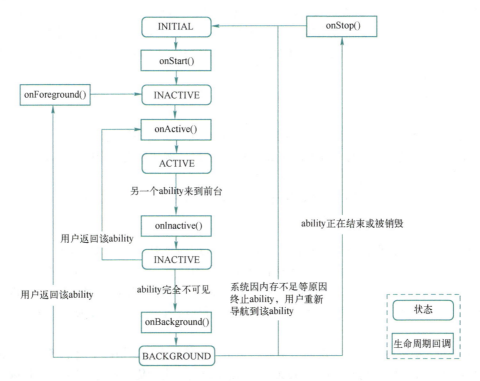

图 14-4　Page 生命周期

（1）onStart()

当系统首次创建 Page 实例时，触发该回调。对于一个 Page 实例，该回调在其生命周期过程中仅触发一次，Page 在该逻辑后将进入 INACTIVE 状态。开发者必须重写该方法，并在此配置默认展示的 AbilitySlice。

（2）onActive()

Page 会在进入 INACTIVE 状态后来到系统前台，然后系统调用此回调。Page 在此之后进入 ACTIVE 状态，该状态是应用与用户交互的状态。Page 将保持在此状态，除非某类事件发生导致 Page 失去焦点，比如用户单击返回键或导航到其他 Page。当此类事件发生时，会触发 Page 回到 INACTIVE 状态，系统将调用 onInactive()回调。此后，Page 可能重新回到 ACTIVE 状态，系统将再次调用 onActive()回调。因此，开发者通常需要成对实现 onActive()和 onInactive()，并在 onActive()中获取在 onInactive()中被释放的资源。

（3）onInactive()

当 Page 失去焦点时，系统将调用此回调，此后 Page 进入 INACTIVE 状态。开发者可以在此回调中实现 Page 失去焦点时应表现的恰当行为。

（4）onBackground()

如果 Page 不再对用户可见，系统将调用此回调通知开发者用户进行相应的资源释放，此后 Page 进入 BACKGROUND 状态。开发者应该在此回调中释放 Page 不可见时无用的资源，或在此回调中执行较为耗时的状态保存操作。

（5）onForeground()

处于 BACKGROUND 状态的 Page 仍然驻留在内存中，当 Page 重新回到系统前台时（比如用户重新导航到此 Page），系统将先调用 onForeground()回调通知开发者，而后 Page 的生命周期状态回到 INACTIVE 状态。开发者应当在此回调中重新申请在 onBackground()中释放的资源，最后 Page 的生命周期状态进一步回到 ACTIVE 状态，系统将通过 onActive()回调通知开发者用户。

（6）onStop()

系统将要销毁 Page 时，将会触发此回调函数，通知用户进行系统资源的释放。销毁 Page 的可能原因包括：

1）用户通过系统管理能力关闭指定 Page，例如使用任务管理器关闭 Page。
2）用户行为触发 Page 的 terminateAbility()方法调用，例如使用应用的退出功能。
3）配置变更导致系统暂时销毁 Page 并重建。
4）系统出于资源管理目的，自动触发对处于 BACKGROUND 状态 Page 的销毁。

2. AbilitySlice 生命周期

AbilitySlice 作为 Page 的组成单元，其生命周期是依托于其所属 Page 生命周期的。AbilitySlice 和 Page 具有相同的生命周期状态和同名的回调，当 Page 生命周期发生变化时，它的 AbilitySlice 也会发生相同的生命周期变化。此外，AbilitySlice 还具有独立于 Page 的生命周期变化，这在同一 Page 中的 AbilitySlice 之间导航时发生，此时 Page 的生命周期状态不会改变。

AbilitySlice 实例的创建和管理通常由应用负责，系统仅在特定情况下会创建 AbilitySlice 实例。例如，通过导航启动某个 AbilitySlice 时，由系统负责实例化；但是在同一个 Page 中不同的 AbilitySlice 间导航时，则由应用负责实例化。

14.3.7 编译运行

编译运行程序，如图 14-5 所示，单击"退出当前账号"，返回登录界面。

图 14-5 返回登录页

14.4 提交代码到仓库

将任务 14 的代码提交到本地仓库，完整流程如下。
1）通过"git status"查看当前目录状态。
2）通过"git add."将当前目录下所有改动加入暂存区。
3）通过"git commit -m 退出当前账号"，提交暂存区内容到本地版本库。
4）通过"git status"查看当前工作目录，可以看到目录干净。
5）通过"git tag -a task14 -m 创建多语言环境"，创建任务 14 标签。
6）通过"git log --pretty=oneline"查看版本库日志，可以看到提交的最新内容，以及 task14 标签，即任务 14 的版本。
7）通过"git tag -l -n"查看所有标签注释。

参 考 文 献

[1] 陈继欣，邓立，谢永华，等. 传感网应用开发职业技能等级标准 [Z]. 北京：北京新大陆时代教育科技有限公司，2019.

[2] SCHILDT H. Java 完全参考手册：第 8 版 [M]. 王德才，吴明飞，唐业军，译. 北京：清华大学出版社，2012.

[3] 华为技术有限公司. HarmonyOS Developer [EB/OL]. [2022-04-07] https://developer.harmonyos.com/.